PHYSICS RESEARCH AND TECHNOLOGY

HARMONIC OSCILLATORS

TYPES, FUNCTIONS AND APPLICATIONS

PHYSICS RESEARCH AND TECHNOLOGY

Additional books and e-books in this series can be found on Nova's website under the Series tab.

PHYSICS RESEARCH AND TECHNOLOGY

HARMONIC OSCILLATORS

TYPES, FUNCTIONS AND APPLICATIONS

YILUN SHANG
EDITOR

nova
science publishers
New York

Library of Congress Cataloging-in-Publication Data

ISBN: 978-1-53615-810-6

Published by Nova Science Publishers, Inc. † New York

CONTENTS

PREFACE

This book gathers state-of-the-art advances on harmonic oscillators including their types, functions, and applications. In Chapter 1, Neetik and Amlan have discussed the recent progresses of information theoretic tools in the context of free and confined harmonic oscillator. Confined quantum systems have provided appreciable interest in areas of physics, chemistry, biology, etc., since its inception. A particle under extreme pressure environment unfolds many fascinating, notable physical and chemical changes. The desired effect is achieved by reducing the spatial boundary from infinity to a finite region. Similarly, in the last decade, information measures were investigated extensively in diverse quantum problems, in both free and constrained situations. The most prominent amongst these are: Fisher information, Shannon entropy, Renyi entropy, Tsallis entropy, Onicescu energy and several complexities. Arguably, these are the most effective measures of uncertainty, as they do not make any reference to some specific points of a respective Hilbert space. These have been invoked to explain several physic-chemical properties of a system under investigation. Kullback Leibler divergence or relative entropy describes how a given probability distribution shifts from a reference distribution function. This characterizes a measure of discrimination between two states. In other words, it extracts the change of information in going from one state to another.

In Chapter 2, Nabakumar, Subhasree, and Paulami have revisited classical-quantum correspondence in the context of linear Simple Harmonic Oscillator (SHO). According to Bohr's correspondence principle, quantum mechanically calculated results match with the classically expected results when quantum number is very high. Classical quantum correspondence may also be visualized in the limit when the action integral is much greater than Planck's constant. When de-Broglie wave length associated with a particle is much larger than system size, then quantum mechanical results also match with the classical results. In the context of dynamics, Ehrenfest equation of motion is used in quantum domain, which is analogous to classical Newton's equation of motion. SHO is one of the most important systems for several reasons. It is one of the few exactly solvable problems. Any stable molecular potential can be approximated by SHO near the equilibrium point. This builds the foundation for the understanding of complex modes of vibration in large molecules, the motion of atoms in a solid lattice, the theory of heat capacity, vibration motion of nuclei in molecule etc. The authors have revisited the common solution techniques and important properties of both classical and quantum linear SHO. Then they focused on probability distribution, quantum mechanical tunneling, classical and quantum dynamics of position, momentum and their actuations, viral theorems, etc. and also analyzed how quantum mechanical results finally tend to classical results in the high quantum number limit.

In Chapter 3, Neeraj has discussed the nature of atomic motions, sometimes referred to as lattice vibrations. The lattice dynamics deals with the vibrations of the atoms inside the crystals. In order to write the dynamic equations of the motion of crystal atoms, we need to describe an inter-atomic interaction. Therefore, it is natural to start the study of the lattice dynamics with the case of small harmonic vibrations. The dynamics of one-dimensional and two-dimensional vibrations of monatomic and diatomic crystals can be understood by using the simple model forces based on harmonic approximation. This harmonic approximation is related to a simple ball-spring model. According to this model, each atom is coupled with the neighboring atoms by spring constants. The collective motion of atoms leads to a distinct traveling wave over the whole crystal, leading to the collective motion, so-called phonon. The simple ball-spring model enlightens us some of the significant common features of lattice dynamics that have been discussed throughout this chapter. Further, this chapter helps in understanding the quantization energy of a harmonic oscillation and the concept of phonon.

In: Harmonic Oscillators
Editor: Yilun Shang
ISBN: 978-1-53615-810-6

Chapter 1

INFORMATION ANALYSIS IN FREE AND CONFINED HARMONIC OSCILLATOR

Neetik Mukherjee and Amlan K. Roy*
Department of Chemical Sciences,
Indian Institute of Science Education and Research Kolkata,
Mohanpur, Nadia, India

Abstract

In this chapter we shall discuss the recent progresses of information theoretic tools in the context of free and confined harmonic oscillator (CHO). Confined quantum systems have provided appreciable interest in areas of physics, chemistry, biology, etc., since its inception. A particle under extreme pressure environment unfolds many fascinating, notable physical and chemical changes. The desired effect is achieved by reducing the spatial boundary from infinity to a finite region. Similarly, in the last decade, information measures were investigated extensively in diverse quantum problems, in both free and constrained situations. The most prominent amongst these are: Fisher information (I), Shannon entropy (S), Rényi entropy (R), Tsallis entropy (T), Onicescu energy (E) and several complexities. Arguably, these are the most effective measures of uncertainty, as they do not make any reference to some specific points of a respective Hilbert space. These have been invoked to explain several physico-chemical properties of a system under investigation. Kullback-Leibler divergence or relative entropy describes how a given probability

*Corresponding Author's E-mail: akroy@iiserkol.ac.in; akroy6k@gmail.com.

distribution shifts from a reference distribution function. This characterizes a measure of discrimination between two states. In other words, it extracts the change of information in going from one state to another.

The one-dimensional confined harmonic oscillator (1DCHO), defined by $v(x) = \frac{1}{2}(x - d_m)^2 + v_c$ (k is the force constant, $v_c = 0,\ x < x_c$ and $v_c = \infty,\ x \geq x_c$), can be classified into two forms, (a) symmetrically confined harmonic oscillator (SCHO) (when $d_m = 0$), (b) asymmetrically confined harmonic oscillator (ACHO) (corresponding to $d_m \neq 0$). Further, in latter case, confinement can be accomplished two different ways: (i) by changing the box boundary, keeping box length and d_m fixed at zero; (ii) another route is to adjust d_m by keeping box length and boundary fixed. SCHO can be treated as an intermediate between a particle-in-a-box (PIB) and a 1DQHO. Though the Schrödinger equation for SCHO can be solved *exactly*, for ACHO it cannot be. We have employed two different methods for the latter, *viz.*, (i) an imaginary time propagation (ITP), leading to minimization of an expectation value (ii) a variation-induced exact diagonalization procedure that utilizes SCHO eigenfunctions as basis. It is found that, at very low x_c region, $I,\ S,\ R,\ T,\ E$ remain invariant with change confinement length, x_c. At moderate x_c region $S_x,\ R_x,\ T_x$ progress and $I_x,\ E_x$ decrease with rise in x_c. Additionally, special attention has been paid to study relative information in SCHO and ACHO.

Analogously, a 3DCHO (radically confined within an impenetrable spherical well) can act as a bridge between particle in a spherical box (PISB) and isotropic 3DQHO. The time-independent Schrödinger equation for D-dimensional harmonic oscillator in both free and confined condition can be solved exactly, within a Dirichlet boundary condition. Here a detailed exploration of information measures has been carried out in r and p spaces, for a 3DCHO. Some exact analytical expressions for these quantities have also been presented, wherever possible. Besides we also consider some recent works on relative information and several complexity measures in composite r and p spaces. Also discussion is made of a recently proposed virial theorem in the context of confinement. In essence, this chapter provides a detailed in-depth investigation about the information theoretical analysis in 1DCHO and 3DCHO, as done in our laboratory.

PACS: 03.65-w, 03.65Ca, 03.65Ta, 03.65.Ge, 03.67-a

Keywords: Rényi entropy, Shannon entropy, Fisher information, Onicescu energy, complexities, confined isotropic harmonic oscillator, particle in a symmetric box

1. INTRODUCTION

"A quantum particle inside an infinite impenetrable box" usually constitutes the first model problem taught in quantum mechanics curriculum. Later, the attention shifts towards the study of "free" systems in some characteristic potential, such as a harmonic oscillator or a H atom in a Coulombic potential, which typify a quantum mechanical problem in a whole region of space. This is itself a recognition of the importance of sub-region Ω of the space, which in this particular scenario is achieved by modeling the boundary condition. Such a situation arises when one tries to explore systems in highly inhomogeneous media or in intense external field. The limit, when the box becomes very small, approaching quantum size is generally termed as "quantum confinement". In this microscopic domain, confinement occurs on a scale comparable to atomic size (object and cavity sizes are commensurate), which is in sharp contrast to the ordinary confinement, where the cavity size is usually much larger than the atom. Classic examples of such macroscopic-size boxes are found in celebrated physics problems like kinetic theory of gas, black-body radiation, atom in a microwave enclosure, etc.

The first seminal work in this direction is due to a pioneering article [1] published in the fourth decade of twentieth century, where the effect of pressure on static polarizability of a H atom enclosed inside a spherical impenetrable hard enclosure was studied. It was realized that certain properties (such as energy/eigenvalue, orbital/eigenfunction, electron localization, shell-filling, polarizability, photonic ionization/absorption, etc., to name a few) of an atom could be significantly altered by "squeezing" or "shrinking" it by placing inside a hard sphere, which acts as an idealized classical piston. Depending on the geometrical form and dimension of cavity, such characteristic features of a caged-in atom exhibits numerous deviations from corresponding *free* atom. Apparently there was some paucity in their immediate applications, but in the last few decades, these models have found wide-spread applications in various branches in physics and chemistry. Lately there has been a plethora of activities in harnessing their potential in various physico-chemical situations as evidenced by an almost exponential growth in the number of articles published in journals, and this continues to grow day by day. Excellent elegant reviews have been made available on the subject; we mention here a very selected set [2, 3, 4, 5, 6].

With the advancement of modern technology, confined quantum systems has emerged as an extremely important contemporary research area, from both theoretical and experimental perspectives. The discovery and development of modern experimental techniques have provided the desired impetus about these boxed-in systems. Constrained atoms, molecules inside various cavities has shed light on their electronic structure, chemical reactivity under high pressure. Apart from that, their relevance has been advocated in numerous other instances, such as impurity in semiconductor/nanostructure, atoms trapped in zeolite sieves or fullerene cage, solutes under solvent environment, artificial atom like quantum dot, quantum wire, quantum well, etc. Recently endohedral metallofullerene clusters where atoms/molecules are typically encapsulated in carbon (or fullerene-like inorganic and hybrid) cages of varying size, shape has found promising applications [7, 8, 9] in engineering. The subtle interaction between host cage and inserted guest molecule is manifested in their marked structural, energetic and spectroscopic properties of endocluster, in comparison to the isolated guest molecule or empty cavity.

The confined harmonic oscillator (CHO) remains one of the oldest, heavily studied model of quantum confinement. In this case the quadratic potential acts through only a finite distance corresponding to the length of box; beyond this an infinite repulsive potential exists. This has been used in the context of various complicated physical situations ever since its inception. Some noteworthy applications include energy generation in dense stars from proton-deuteron transformation [10], mass-radius relation in white dwarf theory [11], rate of escape of a star from a galaxy or globular cumulus [12], electric and magnetic properties of metals including small system of electrons within a cylinder [13, 14], vibronic spectra of point defects, impurities and luminescence in solids [15], specific heats of metal under high pressure, effect of high pressure on properties of materials, etc.

Since the forty's decade, a vast amount of theoretical methods have been reported for accurate solution of a 1DCHO, placed inside a hard impenetrable symmetric box. The usual boundary condition of wave function vanishing at infinity of unenclosed or free harmonic oscillator (QHO) is replaced by the requirement that it vanishes at the walls of enclosure. Eigenfunctions in such systems could then be written in terms of Kummer's function [10, 11], whereas the zeros of confluent hypergeometric function provided energy eigenvalues. Further, the authors employed various approaches to expand the latter in order to find an analytical expression in terms of box size. They found the correct qualitative behavior, i.e., energy levels of a 1DCHO increases rapidly as box length diminishes. An asymptotic expression [16] for energy valid in the region of small length of box was derived. An attempt [17] was made to locate the zeros of hypergeometric function numerically, albeit with limited success. It was further established that while in a QHO (wave function vanishing at ∞), transitions can occur only between two adjacent levels, in a bounded oscillator, there occurs a non-zero transition probability between any two states of different parity. Simple analytical expressions [18] of energy in certain special cases were put forth. The effect of finite boundary on energy levels has also been pursued graphically and from some series expansion [19]. Later, within a semi-classical Wentzel-Kramers-Brillouin (WKB)-type approximation [20], it was pointed out that eigenvalues of the constrained problem reduce to respective unenclosed oscillators, provided the classical turning points remain inside the potential enclosure and not near the walls. Moreover, when the separation of turning points remains large in comparison to the box size, they become plane-wave box eigenvalues. Subsequently, a series analytical solution [21] was proposed for eigenvalues by keeping the center of oscillation at the center of potential enclosure. Eigenvalues were also computed by a numerical procedure as the roots of a polynomial [22]. It was observed that when the box length remains below a certain critical value (corresponding to the effective oscillation length for an unconfined oscillator of lowest energy), the individual energies dramatically increase from uncaged values. Closed-form energy expressions valid for boxes of any size were derived by constructing Padé approximants [23] as interpolation between the perturbative and asymptotic solutions. Methods such as diagonal hyper-virial [24, 25] as well as hyper-virial perturbation theory [26] were also employed for such enclosed systems for a variety of boundary conditions. A Rayleigh-Ritz variation method utilizing the trigonometric basis functions was suggested in [27]. A numerical scheme [28], based on a set of theorems that guarantee that the solution of SE corresponding to a bounded system strongly converges in the norm of Hilbert space $L_2(-\infty, \infty)$ to exact solution of respective unbounded problem. Spatially confined 1DQHO was treated by (WKB) along with modified airy function method [29], as well as supersymmetric WKB approach [30]. Highly accurate eigenvalues were published in [31] by numerically finding zeros of hypergeometric functions. Accurate eigenvalues were computed by power-series method [32], perturbation theory [33], ITP method [34] in conjunction with the minimization of an energy expectation value; in the former also the Einstein coefficients were considered. Very recently, this problem has been revisited by diagonalizing the Hamiltonian using a PIB basis [35]. However, all the above referenced works relate to the so-called *symmetric confinement*; *asymmetric confinement* studies are relatively less. For example, in an asymmetric CHO (ACHO), the energy spectrum have been reported by means of perturbation theory [36], power-series method [32] and ITP [34] method. Other properties such dipole moment [36] as well as Einstein coefficients [36, 32] were also considered.

In parallel to the 1DCHO problem, much attention was also paid on 2D, 3D and N-dimensional counterparts, albeit with less vigour. A number of interesting unique phenomena occur in higher dimensions, especially in connection to *simultaneous, incidental and inter-dimensional degeneracy*, which are elaborated later in due course of time. In one of the earliest attempts, bounded multi-dimensional isotropic harmonic oscillators, including a 3DCHO were investigated using a hyper-virial treatment [37]. Energy eigenvalues of 2DCHO and 3DCHO in circular and spherical boxes respectively were reported through Rayleigh-Schrödinger perturbation expansion, considering the free PIB as corresponding unperturbed systems, as well as Padé approximant solution [38]. A 3DCHO within an impenetrable spherical cavity was approached by a direct variational method [39], where the trial wave function was assumed as product of the "free" solution of corresponding SE and a simple function obeying the respective boundary condition. For the 3DCHO, super-symmetric semi-classical approach [40] has been advocated. The radial SE corresponding to the 2DCHO and 3DCHO was solved quite efficiently by means of a variational procedure, whereby the wave function was expanded into a Fourier-Bessel series [41, 42], with matrix elements involving Bessel functions, that could be evaluated analytically. First detailed and systematic calculation of energies, eigenfunctions and spatial expectation values in isotropic 3DCHO was undertaken in [43], employing a variational strategy. Like the 1DCHO case, a WKB approach was also put forth in 3DCHO by the same author [44], where the centrifugal term is expanded perturbatively (in powers of $\hbar$), partitioning in two parts, *viz*., the usual centrifugal potential governed by classical laws and a quantum correction. A recipe was also provided for 3DCHO within the super-symmetric quantum mechanics [45] in association with variational principle, as well as a generalized pseudo-spectral (GPS) method [46, 47]. Eigenspectra of 2DCHO [48] and 3DCHO [49] were investigated analytically in terms of annihilation and creation operators. A combination of semi-classical WKB method and a proper quantization rule [50] has also been suggested for the spherically confined harmonic oscillator. Some reports have also been made on higher dimensional CHO, although in much lesser intensity. Reasonably [51] and very accurate [52] eigenvalues have been reported for D-dimensional CHO from numerical calculations that find the roots of confluent hypergeometric functions satisfying the necessary degeneracy conditions. High-precision energies were presented [31] by exploiting certain special properties of hypergeometric functions within MAPLE computer algebra system.

In the last few decades, quantum information theory has emerged as a subject of topical interest. This concept has been exploited to understand various phenomena in physics and chemistry, with potential applications in broad topics like thermodynamics [53, 54, 55], quantum mechanics [53, 54, 55], spectroscopy etc. Lately, it has been extensively used to understand quantum entanglement and quantum steering problems [53, 54, 55]. The information theoretic measures like Shannon entropy (S) [56, 57], Rényi entropy (R^α) [58], Fisher information (I), Onicescu energy (E) and complexities (C) are invoked to get knowledge about diffusion of atomic orbitals, spread of electron density, periodic properties, correlation energy and so forth. Perhaps, these are the most effective measures of uncertainty [56, 58, 59, 60], as they do not make any reference to some specific points of the respective Hilbert space. In principle, they can have any real value. However, $(-)$ve value in R and S indicates extreme localization, whereas, E is always $(+)$ve. Likewise, changing the numerical values of R^α, S from $(-)$ve to $(+)$ve only interprets enhancement of de-localization. Another related concept is *complexity*. A system has finite complexity when it is either in a state with less than some maximal order or not at a state of equilibrium. In a nutshell, it becomes *zero* at two limiting cases, *viz.*, when a system is (i) completely ordered (maximum distance from equilibrium) or (ii) at equilibrium (maximum disorder). Complexity has its contemporary interest in chaotic systems, spatial patterns, language, multi-electronic systems, molecular or DNA analysis, social science, astrophysics and cosmology etc.

In this chapter, we summarize the recent progress that has taken place in the information theoretical analysis in case of a quantum harmonic oscillator (QHO) contained in inside a hard, impenetrable cage. To this end, a thorough investigation is made for energy spectra and a host of information quantities, like S, R, I, E, C, in 1DCHO (symmetric and asymmetric) as well as a radially confined QHO. In the 1D case, we introduce two accurate methods for solution of relevant eigenvalue equations in presence of the Dirichlet boundary conditions, namely (i) ITP and (ii) variation-induced exact diagonalization–both developed in our laboratory. They provide accurate eigenvalues and eigenfunctions, which are used throughout. In the 3D counterpart, we employ the very accurate GPS method. One important conclusion is that, a 1DCHO behaves as an intermediate between a PIB and an 1DQHO. Detailed results are compiled in tabular and graphical form for the afore-mentioned information-related quantities, always indicating the transition from bounded to free system. Additionally we also offer a brief account of the *relative information* studies that has been studied by us. Finally, we examine the validity and feasibility of a recently proposed virial theorem in the context of such confined systems, which will consolidate the former's success further. Available literature results have been consulted as and when possible. A few concluding remarks are made at the end.

2. Theoretical Aspects

This section is devoted to the theoretical methods that have been employed for calculation of relevant eigenvalues and eigenfunctions for the spatially confined CHO; both in 1D and 3D. Then we proceed for a discussion on the evaluation of momentum (p)-space wave functions from that in position space, as well as for the computation of information-related quantities.

2.1. Position-Space Wave Function

In 1DCHO problem can be categorized into two distinct forms, *viz.,* (i) in the SCHO case, potential minimum is at the origin ($x = 0$), leading to a *symmetric* confinement (R is length of box), ($v_c(x)$ is the confinement potential), and (ii) ACHO, where the potential minimum is beyond the origin, as exemplified here through the following relations,

$$-\frac{1}{2}\frac{d^2\psi}{dx^2} + \frac{1}{2}(x-d)^2\psi + V(x)\psi = E\psi,$$

where $V(x) = +\infty$, for $|x| \geq R$ and $V(x) = 0$, when $|x| < R$. It represents an infinite square well of width $2R$ with d signifying position of minimum in the potential. The SCHO potential can be solved exactly, whereas in latter case one needs an appropriate numerical method to get the best possible wave function and energy. Like the SCHO case, 3DCHO is also exactly solvable and both wave functions are obtained in the form of *Kummer confluent hypergeometric* function. It is crucial to point out that, in order to construct the exact wave functions of SCHO and 3DCHO for a specific state, however, one requires to provide energy eigenvalue of that state. In our calculation, these energies of SCHO and 3DCHO have been generated from ITP and GPS methods.

2.1.1. Exact Solution

In this part we are going to discuss the exact solutions of SCHO and 3DCHO respectively.

SCHO: The time-independent non-relativistic SE is given by (α is force constant),

$$-\frac{1}{2}\frac{d^2\psi_n(x)}{dx^2} + 4\alpha^2x^2\psi_n(x) + v_c(x)\psi_n(x) = \mathcal{E}_n\psi_n(x), \tag{1}$$

where the confining potential is defined as, $v_c(x) = 0$ for $x < |x_c|$ and $v_c(x) = \infty$ for $x \geq |x_c|$, with x_c signifying length of the impenetrable box. In case of an 1DQHO, $v_c(x) = 0$. The exact analytical solution of Eq. (1) for *even* and *odd* states are as follows,

$$\begin{aligned}
\psi_e(x) &= N_e\ {}_1F_1\left[\left(\frac{1}{4} - \frac{\mathcal{E}_n}{4\sqrt{2}\alpha}\right), \frac{1}{2}, 2\sqrt{2}\alpha x^2\right] e^{-\sqrt{2}\alpha x^2}, \\
\psi_o(x) &= N_o\, x\ {}_1F_1\left[\left(\frac{3}{4} - \frac{\mathcal{E}_n}{4\sqrt{2}\alpha}\right), \frac{3}{2}, 2\sqrt{2}\alpha x^2\right] e^{-\sqrt{2}\alpha x^2}.
\end{aligned} \tag{2}$$

In this equation, N_e, N_o represent normalization constants for even, odd states respectively. It is imperative to mention that, in order to obtain the specific solution for a particular state at a definite x_c, one can adopt either of the following two procedures, *viz.*, (i) use the Dirichlet boundary condition, $\psi_n(-x_c) = \psi_n(x_c) = 0$ at a fixed x_c to obtain $\mathcal{E}_n$ by solving the SE for certain n, and obtain wave function therefrom (ii) provide $\mathcal{E}_n$ as input to get the allowed x_c, which further leads to the wave function. Here, we have solved SE by using ITP procedure at a particular x_c and calculated the respective energy spectrum. Then using this $\mathcal{E}_n$, the desired wave function is constructed to proceed for further calculation. In Eq. (2), ${}_1F_1(a; b; x)$ symbolizes the *Kummer confluent hypergeometric* or confluent hypergeometric function of *1st kind* assuming following form,

$${}_1F_1(a; b; y) = 1 + \frac{a}{b}y + \frac{a(a+1)}{b(b+1)}\frac{y^2}{2!} + \cdots = \sum_{k=0}^{\infty}\frac{a_k}{b_k}\frac{y^k}{k!}. \tag{3}$$

Here a_k, b_k denote the Pochhammer symbols. It is noteworthy to indicate that, if one replaces $\mathcal{E}_n$ by the 1DQHO energy, $(n+\frac{1}{2})2\sqrt{2}\omega$ in Eq. (3), then

$\psi_e(x), \psi_o(x)$ modify as,

$$\begin{aligned}\psi_e(x) = N_e\ {}_1F_1\left[-\left(\frac{j}{2}\right), \frac{1}{2}, 2\sqrt{2}\alpha x^2\right] e^{-\sqrt{2}\alpha x^2} = \\ N_e H_j(2\sqrt{2}\alpha x^2) e^{-\sqrt{2}\alpha x^2}, \\ \psi_o(x) = N_o x\ {}_1F_1\left[-\left(\frac{m}{2}\right), \frac{3}{2}, 2\sqrt{2}\alpha x^2\right] e^{-\sqrt{2}\alpha x^2} = \\ N_o x\ H_m(2\sqrt{2}\alpha x^2) e^{-\sqrt{2}\alpha x^2}.\end{aligned} \tag{4}$$

Here $H_i(2\sqrt{2}\alpha x^2)$ signifies *Hermite* polynomials and j, m correspond to *even, odd* positive integers respectively. Equation (4) clearly suggests that, in an 1DQHO the hypergeometric function reduces to *Hermite* polynomial.

Incidental degeneracy in SCHO: In this instance, quantization is outcome of the boundary condition, which is imposed by making the wave function vanish at x_c. Actually allowed energies are obtained by satisfying ${}_1F_1[a; b; x] = 0$ at x_c. The hypergeometric function takes the form given below, at $x = x_c$ (considering even-parity states),

$$\begin{aligned}{}_1F_1\left[\left(\frac{1}{4} - \frac{\mathcal{E}_n}{4\sqrt{2}\alpha}\right), \frac{1}{2}, 2\sqrt{2}\alpha x_c^2\right] = 1 + \left(\frac{1}{4} - \frac{\mathcal{E}_n}{4\sqrt{2}\alpha}\right) 4\sqrt{2}\alpha x_c^2 + \\ \left(\frac{1}{4} - \frac{\mathcal{E}_n}{4\sqrt{2}\alpha}\right)\left(\frac{5}{4} - \frac{\mathcal{E}_n}{4\sqrt{2}\alpha}\right) \frac{32}{3}\alpha^2 x_c^4 + \\ \left(\frac{1}{4} - \frac{\mathcal{E}_n}{4\sqrt{2}\alpha}\right)\left(\frac{5}{4} - \frac{\mathcal{E}_n}{4\sqrt{2}\alpha}\right)\left(\frac{9}{4} - \frac{\mathcal{E}_n}{4\sqrt{2}\alpha}\right) \frac{128}{15}\alpha^3 x_c^6 + \cdots\end{aligned} \tag{5}$$

Let us consider only two terms in the right-hand series of Eq. (5). Then we get,

$$\left(\frac{5}{4} - \frac{\mathcal{E}_n}{4\sqrt{2}\alpha}\right) = 0, \qquad \mathcal{E}_n = 5\sqrt{2}\alpha. \tag{6}$$

Now, putting the value of $\mathcal{E}_n = 5\sqrt{2}\alpha$ and $x = x_c$ in the boundary condition we get,

$$\begin{aligned} 1 + \left(\frac{1}{4} - \frac{E_n}{4\sqrt{2}\alpha}\right) 4\sqrt{2}\alpha x_c^2 &= 0, \\ 1 + \left(\frac{1}{4} - \frac{5\sqrt{2}\alpha}{4\sqrt{2}\alpha}\right) 4\sqrt{2}\alpha x_c^2 &= 0, \\ (1 - 4\sqrt{2}\alpha x_c^2) &= 0, \\ 4\sqrt{2}\alpha x_c^2 &= 1, \\ x_c &= \pm\frac{1}{2\sqrt{\sqrt{2}\alpha}}. \end{aligned} \tag{7}$$

Since box length is positive, we choose $x_c = \frac{1}{2\sqrt{\sqrt{2}\alpha}}$. This shows that, at this x_c, ${}_1F_1\left[\left(\frac{1}{4} - \frac{\mathcal{E}_n}{4\sqrt{2}\alpha}\right), \frac{1}{2}, 2\sqrt{2}\alpha x^2\right]$ with two terms always represents a non-degenerate ground state having $\mathcal{E}_0 = 5\sqrt{2}\alpha$. Next, let us proceed to the function having three terms. Therefore,

$$\frac{9}{4} - \frac{\mathcal{E}_n}{4\sqrt{2}\alpha} = 0, \qquad \mathcal{E}_n = 9\sqrt{2}\alpha. \tag{8}$$

Using the boundary condition and putting $\mathcal{E}_n = 9\sqrt{2}\alpha$ and $x = x_c$ we obtain,

$$\begin{aligned} 1 + 2\left(\frac{1}{4} - \frac{E_n}{4\sqrt{2}\alpha}\right) 2\sqrt{2}\alpha x_c^2 + \\ \left(\frac{1}{4} - \frac{E_n}{4\sqrt{2}\alpha}\right)\left(\frac{1}{4} - \frac{E_n}{4\sqrt{2}\alpha} + 1\right)\frac{32}{3}\alpha^2 x_c^4 &= 0, \\ 64\alpha^2 x_c^4 - 24\sqrt{2}\alpha x_c^2 + 3 &= 0, \\ x_c^2 = \frac{3+\sqrt{3}}{8\sqrt{2}\alpha} \quad \rightarrow \quad x_c &= \pm\sqrt{\frac{3+\sqrt{3}}{8\sqrt{2}\alpha}}, \\ x_c^2 = \frac{3-\sqrt{3}}{8\sqrt{2}\alpha} \quad \rightarrow \quad x_c &= \pm\sqrt{\frac{3-\sqrt{3}}{8\sqrt{2}\alpha}}. \end{aligned} \tag{9}$$

Equation (9) suggests that, in SCHO, $\mathcal{E}_n = 9\sqrt{2}\alpha$ represents a pair of states; one at $x_c = \sqrt{\frac{3+\sqrt{3}}{8\sqrt{2}\alpha}}$ and other at $x_c = \sqrt{\frac{3-\sqrt{3}}{8\sqrt{2}\alpha}}$. The former represents a second excited state and latter symbolizes ground state. Similar analysis will reveal that, a function ${}_1F_1[a, b, y]$ with $(n+1)$ terms has n number of such degenerate states at n different x_c values. Additionally, in such degenerate set, the smallest box length contains the ground state and longest box length holds the nth state. This procedure helps us to identify incidental degeneracy in SCHO at different box lengths.

3DCHO: The time-independent, non-relativistic wave function for a 3DCHO system, in r space may be written as (n_r identifies radial quantum number),

$$\Psi_{n_r,\ell,m}(\mathbf{r}) = \psi_{n_r,\ell}(r)\, Y_{\ell,m}(\Omega), \tag{10}$$

with r and Ω illustrating radial distance and solid angle successively. Here $\psi_{n_r,\ell}(r)$ represents the radial part and $Y_{\ell,m}(\Omega)$ identifies spherical harmonics of the wave function. The pertinent radial Schrödinger equation under the influence of confinement, with $v(r) = \frac{1}{2}\omega^2 r^2$, is (atomic unit employed unless otherwise mentioned),

$$\left[-\frac{1}{2}\frac{d^2}{dr^2} + \frac{\ell(\ell+1)}{2r^2} + v(r) + v_c(r)\right] \psi_{n_r,\ell}(r) = \mathcal{E}_{n_r,\ell}\, \psi_{n_r,\ell}(r). \tag{11}$$

Our required confinement effect is introduced by invoking the following potential: $v_c(r) = +\infty$ for $r > r_c$, and 0 for $r \leq r_c$, where r_c signifies radius of confinement.

The *exact* generalized radial wave function is mathematically expressed [52] as,

$$\psi_{n_r,\ell}(r) = N_{n_r,\ell}\, r^{\ell}\, {}_1F_1\left[\frac{1}{2}\left(\ell + \frac{3}{2} - \frac{\mathcal{E}_{n_r,\ell}}{\omega}\right), (\ell + \frac{3}{2}), \omega r^2\right] e^{-\frac{\omega}{2}r^2}. \tag{12}$$

Here $N_{n_r,\ell}$ represents normalization constant, $\mathcal{E}_{n_r,\ell}$ corresponds to energy of a given state characterized by quantum numbers n_r, ℓ, whereas ${}_1F_1\,[a, b, r]$ signifies confluent hypergeometric function. Allowed energies are computed by applying the boundary condition $\psi_{n_r,\ell}(0) = \psi_{n_r,\ell}\,(r_c) = 0$. In this work, the GPS method has been used to calculate eigenvalues and eigenfunctions of these states. It has provided highly accurate results for various model and real systems including atoms, molecules, some of which could be found in the references [61, 62, 63, 64, 65, 66, 46, 67, 68, 69, 70, 71, 72, 73, 74, 75, 76, 47, 77, 78].

Incidental degeneracy in 3DCHO: Just like the 1DCHO, quantization in a D-dimensional CHO is also an manifestation of the effect of making radial wave function vanishing at r_c. In this occasion, allowed energies are obtained when,

$$ {}_1F_1\left[\frac{1}{2}\left(\ell+\frac{D}{2}-\frac{\mathcal{E}_{n_r,\ell}}{2\sqrt{2}\alpha}\right);\left(\ell+\frac{D}{2}\right);2\sqrt{2}\alpha r_c^2\right]=0, \tag{13} $$

where, at a fixed ℓ, the successive roots are numbered $n_r = 0, 1, 2, ...$ Note that, the levels are designated by $n_r + 1$ and ℓ values, such that $n_r = \ell = 0$, $n_r = \ell = 2$ correspond to $1s$ and $3d$ states respectively. The radial quantum number n_r relates to n as $n = 2n_r + \ell$. Invoking the series expansion [79] in left-hand side, we get,

$$ \begin{aligned} &{}_1F_1\left[\frac{1}{2}\left(\ell+\frac{D}{2}-\frac{\mathcal{E}_{n_r,\ell}}{2\sqrt{2}\alpha}\right);\left(\ell+\frac{D}{2}\right);2\sqrt{2}\alpha r_c^2\right]= \\ &1+\frac{\left(\frac{\ell}{2}+\frac{D}{4}-\frac{\mathcal{E}_{n_r,\ell}}{4\sqrt{2}\alpha}\right)}{\left(\ell+\frac{D}{2}\right)}2\sqrt{2}\alpha r_c^2+ \\ &\frac{\left(\frac{\ell}{2}+\frac{D}{4}-\frac{\mathcal{E}_{n_r,\ell}}{4\sqrt{2}\alpha}\right)\left(\frac{\ell}{2}+\frac{D}{4}-\frac{\mathcal{E}_{n_r,\ell}}{4\sqrt{2}\alpha}+1\right)}{\left(\ell+\frac{D}{2}\right)\left(\ell+\frac{D}{2}+1\right)}\frac{8}{2!}\alpha^2 r_c^4+\cdots \end{aligned} \tag{14} $$

In order to terminate the series after 2nd term we need,

$$ \begin{aligned} \left(\frac{\ell}{2}+\frac{D}{4}-\frac{\mathcal{E}_{n_r,\ell}}{4\sqrt{2}\alpha}+1\right)&=0 \\ \frac{\mathcal{E}_{n_r,\ell}}{2\sqrt{2}\alpha}&=1+\frac{\ell}{2}+\frac{D}{4} \\ \mathcal{E}_{n_r,\ell}&=\sqrt{2}\alpha\left(D+2\ell+4\right). \end{aligned} \tag{15} $$

Now, it is important to determine the corresponding r_c values at which Eq. (15) is valid. According to Eq. (14), this corresponds to the requirement that,

$$ \begin{aligned} 1+\frac{\left(\frac{\ell}{2}+\frac{D}{4}-\frac{\mathcal{E}_{n_r,\ell}}{4\sqrt{2}\alpha}\right)}{\left(\ell+\frac{D}{2}\right)}2\sqrt{2}\alpha r_c^2&=0 \\ 1+\frac{\left(\frac{\ell}{2}+\frac{D}{4}-\frac{\sqrt{2}\alpha(D+2\ell+4)}{4\sqrt{2}\alpha}\right)}{\left(\ell+\frac{D}{2}\right)}2\sqrt{2}\alpha r_c^2&=0 \\ r_c&=\sqrt{\frac{1}{2\sqrt{2}\alpha}\left(\ell+\frac{D}{2}\right)}. \end{aligned} \tag{16} $$

This shows that, degeneracy depends on ℓ and D values. Let us consider some examples,

1. $l = 1,\ D = 2,\ \mathcal{E}_{n_r,1} = 8\sqrt{2}\alpha, \quad l = 0,\ D = 4,\ \mathcal{E}_{n_r,0} = 8\sqrt{2}\alpha$ at $r_c = \sqrt{\frac{1}{\sqrt{2}\alpha}}$,

2. $l = 1,\ D = 3,\ \mathcal{E}_{n_r,1} = 9\sqrt{2}\alpha, \quad l = 0,\ D = 5,\ \mathcal{E}_{n_r,0} = 9\sqrt{2}\alpha$ at $r_c = \sqrt{\frac{5}{4\sqrt{2}\alpha}}$,

3. $l = 1,\ D = 2,\ \mathcal{E}_{n_r,2} = 10\sqrt{2}\alpha, \quad l = 2,\ D = 4,\ \mathcal{E}_{n_r,1} = 10\sqrt{2}\alpha$ at $r_c = \sqrt{\frac{3}{2\sqrt{2}\alpha}}$,

4. $l = 1,\ D = 3,\ \mathcal{E}_{n_r,2} = 11\sqrt{2}\alpha, \quad l = 2,\ D = 5,\ \mathcal{E}_{n_r,1} = 11\sqrt{2}\alpha$ at $r_c = \sqrt{\frac{7}{4\sqrt{2}\alpha}}$,

5. $l = 2,\ D = 4,\ \mathcal{E}_{n_r,2} = 12\sqrt{2}\alpha, \quad l = 3,\ D = 2,\ \mathcal{E}_{n_r,1} = 12\sqrt{2}\alpha$ at $r_c = \sqrt{\frac{2}{\sqrt{2}\alpha}}$.

From above instances it is clear that, degeneracy appears at a certain r_c. However, the selection rule for this degeneracy can be achieved from the following relation. Suppose, at a fixed r_c we have a pair of states, characterized by quantum numbers l_1, l_2, in dimensions D_1, D_2 respectively, having same energy. As a consequence of the above, we have,

$$\sqrt{2}\alpha(D_1 + 2l_1 + 4) = \sqrt{2}\alpha(D_2 + 2l_2 + 4). \tag{17}$$

$$\sqrt{\frac{1}{2\sqrt{2}\alpha}\left(l_1 + \frac{D_1}{2}\right)} = \sqrt{\frac{1}{2\sqrt{2}\alpha}\left(l_2 + \frac{D_2}{2}\right)} \tag{18}$$

Equations (17) and (18) lead to the common relation,

$$(D_2 - D_1) = 2(l_1 - l_2). \tag{19}$$

It is noticeable from Eq. (19) that, if $D_2 > D_1$ then $l_1 > l_2$ and vice-versa. Thus the incidental degeneracy selection rule is,

$$(n_r, l, D, r_c) \quad \rightarrow \quad (n_r, l \pm j, D \mp 2j, r_c), \tag{20}$$

where j is an integer. It is noticed that, Eq. (20) may further be written in following form,

$$\begin{aligned} \mathcal{E}^{D}_{n_r,(l+j)}(r_c) &= \mathcal{E}^{(D+2j)}_{n_r,l}(r_c), \\ \mathcal{E}^{(D-2j)}_{n_r,l}(r_c) &= \mathcal{E}^{D}_{n_r,(l-j)}(r_c). \end{aligned} \tag{21}$$

Thus, in a 3DCHO, incidental and inter-dimensional degeneracy occur simultaneously.

Simultaneous degeneracy in 3DCHO: It can be proved that, at $r_c = \sqrt{\frac{1}{2\sqrt{2}\alpha}\left(l+\frac{D}{2}\right)}$, the pair of energies $\mathcal{E}^{D}_{n_r,(l+2)}$ and $\mathcal{E}^{D}_{(n_r+1),l}$ are related by the following relation,

$$\mathcal{E}^{D}_{(n_r+1),l}\left(\sqrt{\frac{1}{2\sqrt{2}\alpha}\left(l+\frac{D}{2}\right)}\right) = \mathcal{E}^{D}_{n_r,(l+2)}\left(\sqrt{\frac{1}{2\sqrt{2}\alpha}\left(l+\frac{D}{2}\right)}\right). \tag{22}$$

Inter-dimensional degeneracy in 3DCHO: Such a degeneracy occurs when,

$$(n_r, l, D) \rightarrow (n_r, l \pm j, D \mp 2j). \tag{23}$$

This can be demonstrated by the example as given below,

$$\mathcal{E}^{2}_{0,4}(r_c) = \mathcal{E}^{4}_{0,3}(r_c) = \mathcal{E}^{6}_{0,2}(r_c) = \mathcal{E}^{8}_{0,1}(r_c) \tag{24}$$

In this scenario, the terms inter-dimensional and incidental degeneracy are synonymous.

2.1.2. Imaginary-Time Propagation (ITP) Method

In this subsection, we discuss the general concepts of the imaginary-time evolution technique, as employed here for a particle under confinement. The scheme has been found to provide accurate bound-state solutions through a transformation of the appropriate time-dependent (TD) SE into a diffusion equation in imaginary time. This resulting equation is numerically solved through a finite-difference procedure, in amalgamation with a minimization of expectation value to hit the ground state. After the original proposal that came several decades ago, a number of successful implementations [80, 81, 82, 83, 84, 85, 86] have been reported in the literature since then. In this work we have adopted an implementation, which has been successfully applied to a number of physical systems, such as atoms, diatomic molecules within a quantum fluid dynamical density functional theory (DFT) [87, 88], as well as some model (harmonic, anharmonic, self-interacting, double-well, spiked oscillators) potentials [89, 90, 91, 92, 93], in both 1D, 2D and 3D. Recently this was also extended to confinement (SCHO and ACHO) problems [34] with very good success.

In the following, we provide a short account of the essentials of the methodology; more details could be found in the references quoted above. Let us begin with the TDSE, which for a particle under the influence of a potential $v(x)$ in 1D, in atomic unit, is given by ($v(x) = 4\omega^2 x^2$ in an SCHO),

$$i\frac{\partial}{\partial t}\psi(x,t) = H\psi(x,t) = \left[-\frac{1}{2}\frac{d^2}{dx^2} + v(x) + v_c(x)\right]\psi(x,t). \quad (25)$$

The Hamiltonian operator consists of usual kinetic and potential energy operators. This method is, *in principle, exact*. Here we have provided the equations for 1D SCHO; however this has been easily extended to higher dimensions (see, e.g., [92, 93]. The confinement condition can be achieved by reducing the boundary from *infinity* to finite region, as expressed in the following equation (symmetric box of length $2R$),

$$v_c(x) = \begin{cases} 0, & -R < x < +R \\ +\infty, & |x| \geq R. \end{cases} \quad (26)$$

Equation (25) can be expressed in imaginary time, $\tau = it$ (t is real time) to yield a non-linear diffusion-type equation similar to a diffusion quantum Monte Carlo equation [94],

$$-\frac{d\psi(x,\tau)}{d\tau} = H\psi(x,\tau). \quad (27)$$

Its formal solution can be written as,

$$\psi(x,\tau) = \sum_{k=0}^{\infty} c_k \psi_k(x) \exp\left(-\epsilon_k \tau\right). \quad (28)$$

The initial guessed wave function has the form, $\psi(x,\tau)$ at $\tau = 0$ and if propagated for a sufficiently long time, it will finally converge towards the desired stationary ground-state wave function; $\lim_{\tau\to\infty}\psi(x,\tau) \approx c_0\psi_0(x)e^{-\epsilon_0\tau}$. Thus, provided $c_0 \neq 0$, apart from a normalization constant, this leads to the global minimum corresponding to an expectation value $\langle\psi(x,\tau)|H|\psi(x,\tau)\rangle$.

The numerical solution of Eq. (27) can be obtained by using a Taylor series expansion of $\psi(x, \tau + \Delta\tau)$ around time τ as follows,

$$\psi(x, \tau + \Delta t) = e^{-\Delta\tau H}\psi(x, \tau). \tag{29}$$

Here, the exponential in right side represents a time-evolution operator, which propagates diffusion function $\psi(x, \tau)$ at an initial time τ to an advanced time step to $\psi(x, \tau + \Delta\tau)$. Since this is a non-unitary operator, normalization of the function at a given time τ does not necessarily preserve the same at a future time $\tau + \Delta\tau$. Transformation of Eq. (29) into an equivalent, symmetrical form leads to (j, n signify space, time indices respectively),

$$e^{(\Delta\tau/2)H_j}\ \psi_j^{'(n+1)} = e^{-(\Delta\tau/2)H_j}\ \psi_j^n. \tag{30}$$

A prime is introduced in above equation to indicate the *unnormalized* diffusion function. Taking the full form of Hamiltonian from Eq.(25), one can further write,

$$e^{(\Delta\tau/2)[-\frac{1}{2}D_x^2+v(x_j)]}\ \psi_j^{'(n+1)} = e^{-(\Delta\tau/2)[-\frac{1}{2}D_x^2+v(x_j)]}\ \psi_j^n, \tag{31}$$

where the spatial second derivative has been defined as $D_x^2 = \frac{d^2}{dx^2}$. Now, expanding the exponentials on both sides, followed by truncation after second term and approximation of second derivative by a five-point difference formula [79] ($\Delta x = h$),

$$D_x^2\ \psi_j^n \approx \frac{-\psi_{j-2}^n + 16\psi_{j-1}^n - 30\psi_j^n + 16\psi_{j+1}^n - \psi_{j+2}^n}{12h^2}, \tag{32}$$

yields a set of N simultaneous equations, as follows,

$$\alpha_j\psi_{j-2}^{'(n+1)} + \beta_j\psi_{j-1}^{'(n+1)} + \gamma_j\psi_j^{'(n+1)}\delta_j\psi_{j+1}^{'(n+1)} + \zeta_j\psi_{j+2}^{'(n+1)} = \xi_j^n. \tag{33}$$

After some straightforward algebra, the quantities $\alpha_j, \beta_j, \gamma_j, \delta_j, \zeta_j, \xi_j^n$ are identified as,

$$\begin{aligned}
\alpha_j &= \zeta_j = \frac{\Delta\tau}{48h^2},\ \beta_j = \delta_j = -\frac{\Delta\tau}{3h^2},\ \ \gamma_j = 1 + \frac{5\Delta\tau}{8h^2} + \frac{\Delta\tau}{2}\,v(x_j) \quad (34)\\
\xi_j^n &= \left[-\frac{\Delta\tau}{48h^2}\right]\psi_{j-2}^n + \left[\frac{\Delta\tau}{3h^2}\right]\psi_{j-1}^n + \left[1 - \frac{5\Delta\tau}{8h^2} - \frac{\Delta\tau}{2}\,v(x)\right]\psi_j^n \\
&\quad + \left[\frac{\Delta\tau}{3h^2}\right]\psi_{j+1}^n - \left[-\frac{\Delta\tau}{48h^2}\right]\psi_{j+2}^n.
\end{aligned}$$

Since discretization and truncation occur on both sides, there may be some cancellation of errors. Here, $\psi'^{(n+1)}$ signifies the unnormalized diffusion function at some future time τ_{n+1}. The quantities like $\alpha_j, \beta_j, \gamma_j, \delta_j, \zeta_j$, after some algebraic manipulation, can be easily expressed in terms of space and time spacings as well as the potential (appears only in γ_j and ξ_j^n). The latter also requires knowledge of normalized diffusion functions $\psi_{j-2}^n, \psi_{j-1}^n, \psi_j^n, \psi_{j+1}^n, \psi_{j+2}^n$ at spatial grid points $x_{j-2}, x_{j-1}, x_j, x_{j+1}, x_{j+2}$ at an earlier time τ_n. It is convenient to recast this in a pentadiagonal matrix form, as below,

$$\begin{bmatrix} \gamma_1 & \delta_1 & \zeta_1 & & & & (0) \\ \beta_2 & \gamma_2 & \delta_2 & \zeta_2 & & & \\ \alpha_3 & \beta_3 & \gamma_3 & \delta_3 & \zeta_3 & & \\ & \ddots & \ddots & \ddots & \ddots & \ddots & \\ & & \ddots & \ddots & \ddots & \ddots & \zeta_{N-2} \\ & & & \alpha_{N-1} & \beta_{N-1} & \gamma_{N-1} & \delta_{N-1} \\ (0) & & & & \alpha_N & \beta_N & \gamma_N \end{bmatrix} \begin{bmatrix} \psi_1'^{(n+1)} \\ \psi_2'^{(n+1)} \\ \psi_3'^{(n+1)} \\ \vdots \\ \psi_{N-2}'^{(n+1)} \\ \psi_{N-1}'^{(n+1)} \\ \psi_N'^{(n+1)} \end{bmatrix} = \begin{bmatrix} \xi_1^n \\ \xi_2^n \\ \xi_3^n \\ \vdots \\ \xi_{N-2}^n \\ \xi_{N-1}^n \\ \xi_N^n \end{bmatrix}. \tag{35}$$

The above matrix equation can be readily solved for $\{\psi'^{(n+1)}\}$ using standard routine, satisfying the required boundary condition $\psi_1^n = \psi_N^n = 0$, at all time. Hence, from an initial trial function ψ_j^0 at $n = 0$ time step, the diffusion function can be propagated according to Eq. (29) following the sequence of steps as delineated above. Then at a given time level $(n+1)$, the following series of instructions are carried out, *viz.*, (a) $\psi_j'^{(n+1)}$ is normalized to $\psi_j^{(n+1)}$ (b) if one is interested in an excited-state calculation, then $\psi_j'^{(n+1)}$ is required to be made orthogonalized with respect to all lower states (we use the standard Gram-Schmidt method) (c) desired expectation values are computed as $\epsilon_0 = \langle \psi^{(n+1)} | H | \psi^{(n+1)} \rangle$ (d) difference in the observable expectation values between two successive time steps, $\Delta\epsilon = \langle H \rangle^{(n+1)} - \langle H \rangle^n$, is monitored, and (e) until the above discrepancy $\Delta\epsilon$, goes below a certain threshold tolerance limit, one proceeds with the calculation of $\psi_j^{(n+2)}$ at the next time level iteratively. The guessed functions for even and odd states were chosen as simple Gaussian-type functions such as e^{-x^2} and xe^{-x^2} respectively. All the integrals were evaluated with the help of standard Newton-Cotes quadratures [79].

2.1.3. Generalized Pseudospectral (GPS) Method

All the 3DCHO calculations in this work, were done by using the the GPS formalism. It provides accurate eigenvalues and eigenfunctions easily various r_c. By means of an optimal, non-uniform spatial grid, it leads to a symmetric eigenvalue problem, which can be easily solved by standard diagonalization routines available. It has been applied to a series of model and real systems, in both *free and confined* cases, *viz.*, spiked harmonic oscillators [61, 70], power-law and logarithmic [62], Húlthen and Yukawa [64], Hellmann [68], rational [69], exponential-screened Coulomb [71], Morse [72], hyperbolic [73], Deng-Fan [74], Manning-Rosen [75], Tietz-Hua [76], other singular [65] potentials, as well as many-electron systems within the broad domain of DFT [95, 96, 88, 63, 66, 67]. Of late this has also produced excellent quality results in various radial confinement [46, 47, 77, 78] studies in several Coulombic systems, as well as in 3DCHO.

The key characteristic of this method is to approximate an *exact* wave function $f(x)$ defined in the period $[-1, 1]$ by the Nth-order polynomial $f_N(x)$.

$$f(x) \cong f_N(x) = \sum_{j=0}^{N} f(x_j) g_j(x), \tag{36}$$

and confirm the estimation to be *exact* at the *collocation points* x_j,

$$f_N(x_j) = f(x_j). \tag{37}$$

In this work, the authors have employed the Legendre pseudo-spectral method where $x_0 = -1, x_N = 1$, and $x_j (j = 1, ..., N-1)$ are defined by the roots of first derivative of Legendre polynomial $P_N(x)$, with respect to x, namely,

$$P_N^{'}(x_j) = 0. \tag{38}$$

In Eq. (36), $g_j(x)$ are termed *cardinal functions*, and as such, are given by.

$$g_j(x) = -\frac{1}{N(N+1)P_N(x_j)} \frac{(1-x^2)P_N^{'}(x)}{(x-x_j)}, \tag{39}$$

fulfilling the unique property that, $g_j(x_{j'}) = \delta_{j',j}$.

Table 1. The co-efficients a_k, b_j for even-l p-space wave functions in central potential [97]

l	b_0	b_2	b_4	b_6	b_8	a_1	a_3	a_5	a_7
0	$\frac{1}{\sqrt{\pi}}$	–	–	–	–	–	–	–	–
2	$\frac{1}{\sqrt{\pi}}$	$-\frac{3}{\sqrt{\pi}}$	–	–	–	$\frac{3}{\sqrt{\pi}}$	–	–	–
4	$\frac{1}{\sqrt{\pi}}$	$-\frac{105}{\sqrt{\pi}}$	$\frac{315}{\sqrt{\pi}}$	–	–	$\frac{30}{\sqrt{\pi}}$	$-\frac{315}{\sqrt{\pi}}$	–	–
6	$\frac{1}{\sqrt{\pi}}$	$-\frac{210}{\sqrt{\pi}}$	$\frac{4725}{\sqrt{\pi}}$	$-\frac{10395}{\sqrt{\pi}}$	–	$\frac{21}{\sqrt{\pi}}$	$\frac{-1260}{\sqrt{\pi}}$	$\frac{10395}{\sqrt{\pi}}$	–
8	$\frac{1}{\sqrt{\pi}}$	$-\frac{630}{\sqrt{\pi}}$	$\frac{51975}{\sqrt{\pi}}$	$-\frac{945945}{\sqrt{\pi}}$	$\frac{2027025}{\sqrt{\pi}}$	$\frac{36}{\sqrt{\pi}}$	$-\frac{6930}{\sqrt{\pi}}$	$\frac{270270}{\sqrt{\pi}}$	$-\frac{2027025}{\sqrt{\pi}}$

In order to solve the radial SE for a central potential using finite-difference methods, it requires a significantly large number of points in an equal-spacing grid arrangement. However, in GPS method, this is alleviated by (i) mapping semi-infinite domain $r \in [0, \infty]$ onto a finite domain $x \in [-1, 1]$ via a transformation $r = r(x)$, and then (ii) employing a Legendre pseudo-spectral discretization technique. Utilizing an algebraic non-linear mapping,

$$r = r(x) = L\frac{(1+x)}{(1-x+\alpha)}, \tag{40}$$

where $L, \alpha = \frac{2L}{r_{max}}$ denote two mapping parameters, plus a symmetrization procedure,

$$\psi(r(x)) = \sqrt{r'(x)}f(x), \tag{41}$$

leads to the transformed Hamiltonian, as given below,

$$\hat{H}(x) = -\frac{1}{2}\frac{1}{r'(x)}\frac{d^2}{dx^2}\frac{1}{r'(x)} + V(r(x)) + V_m(x), \tag{42}$$

where,

$$V_m(x) = \frac{3(r'')^2 - 2r'''r'}{8(r')^4}. \tag{43}$$

This turns out to be a symmetric matrix eigenvalue problem. The mapping used in Eq. (40) is such that $V_m(x) = 0$. Eventually, we obtain following discrete set

Table 2. The co-efficients a_k, b_j for odd-l p-space wave functions in central potential [97]

l	a_0	a_2	a_4	a_6	a_8
1	$\frac{1}{\sqrt{\pi}}$	–	–	–	–
3	$\frac{1}{\sqrt{\pi}}$	$-\frac{15}{\sqrt{\pi}}$	–	–	–
5	$\frac{1}{\sqrt{\pi}}$	$-\frac{105}{\sqrt{\pi}}$	$\frac{945}{\sqrt{\pi}}$	–	–
7	$\frac{1}{\sqrt{\pi}}$	$-\frac{378}{\sqrt{\pi}}$	$\frac{17325}{\sqrt{\pi}}$	$-\frac{135135}{\sqrt{\pi}}$	–
9	$\frac{1}{\sqrt{\pi}}$	$-\frac{990}{\sqrt{\pi}}$	$\frac{135135}{\sqrt{\pi}}$	$-\frac{4729725}{\sqrt{\pi}}$	$\frac{34459425}{\sqrt{\pi}}$
l	b_1	b_3	b_5	b_7	b_9
1	$-\frac{1}{\sqrt{\pi}}$	–	–	–	–
3	$-\frac{3!}{\sqrt{\pi}}$	$\frac{15}{\sqrt{\pi}}$	–	–	–
5	$-\frac{15}{\sqrt{\pi}}$	$\frac{420}{\sqrt{\pi}}$	$-\frac{945}{\sqrt{\pi}}$	–	–
7	$-\frac{28}{\sqrt{\pi}}$	$\frac{3150}{\sqrt{\pi}}$	$-\frac{2370}{\sqrt{\pi}}$	$\frac{135135}{\sqrt{\pi}}$	–
9	$-\frac{45}{\sqrt{\pi}}$	$\frac{13860}{\sqrt{\pi}}$	$-\frac{945945}{\sqrt{\pi}}$	$\frac{16216200}{\sqrt{\pi}}$	$-\frac{34459425}{\sqrt{\pi}}$

of coupled equations,

$$\sum_{j=0}^{N}\left[-\frac{1}{2}D^{(2)}_{j'j} + \delta_{j'j}V(r(x_j)) + \delta_{j'j}V_m(r(x_j))\right] A_j = \mathrm{E}A_{j'},\ J = 1, ..., N-1, \tag{44}$$

$$A_j = R'(x_j)f(x_j)[P_N(x_j)]^{-1} = [r'(x_j)]^{\frac{1}{2}}\psi(r(x_j))[P_N(x_j)]^{-1}. \tag{45}$$

Here, $D^{(2)}_{j'j}$ signifies symmetrized second derivative of cardinal function with respect to r,

$$D^{(2)}_{j'j} = [r'(x_{j'})]^{-1} d^{(2)}_{j'j} [r'(x_j)]^{-1}, \tag{46}$$

$$d^{(2)}_{j'j} = \begin{cases} \frac{(N+1)(N+2)}{6(1-x_j^2)}, & j = j', \\ \frac{1}{(x_j - x_{j'})^2}, & j \neq j'. \end{cases} \tag{47}$$

The symmetric eigenvalue problem can be easily and efficiently be solved by standard library routines, to obtain accurate eigenvalues and eigenfunctions.

2.2. Momentum-Space Wave Function

The p-space wave function ($\mathbf{p} = \{p, \Omega\}$) for a particle in a central potential is obtained from respective Fourier transform of its r-space counterpart [97], and as such, is given below,

$$\begin{aligned}\psi_{n,l}(p) &= \frac{1}{(2\pi)^{\frac{3}{2}}} \int_0^\infty \int_0^\pi \int_0^{2\pi} \psi_{n,l}(r)\ \Theta(\theta)\Phi(\phi)\ e^{ipr\cos\theta} r^2 \sin\theta\ \mathrm{d}r\mathrm{d}\theta\mathrm{d}\phi, \\ &= \frac{1}{2\pi}\sqrt{\frac{2l+1}{2}} \int_0^\infty \int_0^\pi \psi_{n,l}(r)\ P_l^0(\cos\theta)\ e^{ipr\cos\theta}\ r^2 \sin\theta\ \mathrm{d}r\mathrm{d}\theta.\end{aligned} \tag{48}$$

Note that $\psi(p)$ is not normalized; thus needs to be normalized. Integrating over θ and ϕ variables, Eq. (48) can be further reduced to,

$$\psi_{n,l}(p) = (-i)^l \int_0^\infty \frac{\psi_{n,l}(r)}{p}\ f(r,p)\mathrm{d}r. \tag{49}$$

Depending on l, this can be rewritten in following simplified form (m' starts with 0),

$$\begin{aligned} f(r,p) &= \sum_{k=2m'+1}^{m'<\frac{l}{2}} a_k \frac{\cos pr}{p^k r^{k-1}} + \sum_{j=2m'}^{m'=\frac{l}{2}} b_j \frac{\sin pr}{p^j r^{j-1}}, \quad \text{for even } l, \\ f(r,p) &= \sum_{k=2m'}^{m'=\frac{l-1}{2}} a_k \frac{\cos pr}{p^k r^{k-1}} + \sum_{j=2m'+1}^{m'=\frac{l-1}{2}} b_j \frac{\sin pr}{p^j r^{j-1}}, \quad \text{for odd } l. \end{aligned} \tag{50}$$

The co-efficients a_k, b_j of even- and odd-l states are collected in Tables 1 and 2 respectively.

3. Formulation of Information-Theoretical Quantities

In this section we shall briefly discuss the various information-theoretic quantities along with their mathematical form.s This will provide the context where these quantities are defined and the relations between them.

3.1. Shannon Entropy (S)

Information is carried over from one place to another. We seek information when there are more than one alternatives, and we are not certain about the outcome of the event. If an event occurs in just one way, there is no uncertainty and no information is called for. In summary, we get some information due to the occurrence of an event if there was some uncertainty prevailing before the occurrence [56, 98, 99].

$$\text{Information received by occurrence of an event} = \text{Amount of uncertainty prevailed before its occurrence} \tag{51}$$

Let us consider a discrete probability distribution $(p_1, p_2, p_3, ..., p_n)$ consisting of n different events. To quantify the uncertainty in results, Shannon in 1948 defined the information entropy as [57],

$$S = -k \sum p_i \ln p_i, \tag{52}$$

where k is a positive constant depending on the choice of unit. This definition can be explained by choosing two limiting cases: (i) at first, when any of the $p_i = 1$ and others are *zero*, then for this certain event $S = 0$, which is minimum (ii) in case of equiprobability, where all $p_i = \frac{1}{n}$, and the uncertainty of the outcome is maximum, then S is also maximum ($S = k \ln n$) [98]. In essence, it can be said that, for a given distribution, lesser the probability of occurring an event, higher will be the uncertainty associated with it. Hence, after occurrence of that event, more information will come out. Arguably, S is the best measure of information [56].

In wave mechanics, the concept of S has been used to explain and interpret various phenomena. Illustrative examples involve in illuminating Colin conjecture [100, 101], atomic avoided crossing [102, 103], orbital-free DFT [104, 105], electron correlation [106, 107, 108, 109, 110, 111], configuration-interaction, entanglement in artificial atoms [112], aromaticity [113] in many-electron systems, etc. Moreover, it can be proved that a stronger version (compared to the traditional position-momentum uncertainty, due to Heisenberg) has been derived by adopting the idea of S [56],

$$S_{\mathbf{r}} + S_{\mathbf{p}} \geq D(1 + \ln \pi), \tag{53}$$

where D refers to dimensionality of the system. Here, $S_{\mathbf{r}}$, $S_{\mathbf{p}}$ have the form,

$$\begin{aligned}
S_{\mathbf{r}} &= -\int_{\mathcal{R}^3} \rho(\mathbf{r}) \ln[\rho(\mathbf{r})] \, \mathrm{d}\mathbf{r} = 2\pi \left(S_r + S_{(\theta,\phi)}\right), \\
S_{\mathbf{p}} &= -\int_{\mathcal{R}^3} \Pi(\mathbf{p}) \ln[\Pi(\mathbf{p})] \, \mathrm{d}\mathbf{p} = 2\pi \left(S_p + S_{(\theta,\phi)}\right), \\
S_r &= -\int_0^\infty \rho(r) \ln[\rho(r)] r^2 \mathrm{d}r, \qquad \rho(r) = |\psi_{n,l}(r)|^2, \\
S_p &= -\int_0^\infty \Pi(p) \ln[\Pi(p)] \, p^2 \mathrm{d}p, \qquad \Pi(p) = |\psi_{n,l}(p)|^2, \\
S_{(\theta,\phi)} &= -\int_0^\pi \chi(\theta) \ln[\chi(\theta)] \sin\theta \mathrm{d}\theta, \qquad \chi(\theta) = |\Theta(\theta)|^2.
\end{aligned} \tag{54}$$

3.2. Rényi Entropy (R)

Rényi entropy is a one-parameter extension of S. Rényi in 1961 defined this quantity as, "the measure of information of order α associated with the probability distribution $P = (p_1, p_2, ..., p_n)$" [114]. The mathematical form of R is,

$$R_\alpha = \frac{1}{1-\alpha} \ln \left(\sum_k p_k^\alpha \right). \tag{55}$$

It is important to note that, R_α is considered as another measure of information because (i) R_α is the exponential mean of information entropy whereas S provides the arithmetic mean of it, and (ii) at $\alpha \to \infty$, R_α reduces to S [56].

It is well known that, R^α, the so-called information generating functional, is closely related to α order entropic moments, and completely characterize density $\rho(\mathbf{r})$. In case of continuous density distribution, it is expressed in terms of expectation values of density, in the following conventional form [114, 115],

$$R^\alpha[\rho(\mathbf{r})] = \frac{1}{(1-\alpha)} \ln \langle \rho(\mathbf{r})^{(\alpha-1)} \rangle, \quad 0 < \alpha < \infty, \ \alpha \neq 1. \tag{56}$$

Quantum mechanical uncertainty principle such as, $\Delta x \Delta p \geq \hbar$, does not comment about the accuracy of our measuring instruments. On the contrary, entropic uncertainty relations depend on the accuracy of a given measurement. Because they manifest the area of phase-space ($\delta\mathbf{r}\delta\mathbf{p}$) obtained by the resolution of measuring instruments. It suggests that, with an increase of localization of a particle in phase space, the sum of uncertainties in position and momentum space enhances. The quantum mechanical uncertainty relation containing the phase space is as follows,

$$R_{\mathbf{r}}^{\alpha} + R_{\mathbf{p}}^{\alpha} \geq -\frac{1}{2}\left(\frac{\ln \alpha}{1-\alpha} + \frac{\ln \beta}{1-\beta}\right) - \ln \frac{\delta \mathbf{r} \delta \mathbf{p}}{\pi \hbar}; \quad \left(\frac{1}{\alpha} + \frac{1}{\beta}\right) = 2. \tag{57}$$

In the limit, when $\alpha \rightarrow 1$ and $\beta \rightarrow 1$, this uncertainty relation reduces to the relation given in Eq. (53). The relation given in Eq. (57) provides better idea about uncertainty as it contains all-order entropic moments [58]. But its improvement is still necessary, which remains an open challenging problem. Interestingly, Eq. (57) becomes sharper and sharper when the relative size of phase-space area $\left(\frac{\delta \mathbf{r} \delta \mathbf{p}}{\pi \hbar}\right)$ defined by the experimental resolution decreases; as it is when we enter deeper and deeper into the quantum regime.

In quantum mechanics, R^{α} has been successfully employed to investigate and predict various quantum properties and phenomena like entanglement, communication protocol, correlation de-coherence, measurement, localization properties of Rydberg states, molecular reactivity, multi-fractal thermodynamics, production of multi-particle in high-energy collision, disordered systems, spin system, quantum-classical correspondence, localization in phase space [116, 117, 118, 119, 120, 121, 122, 123, 124, 125], etc.

Rényi entropies of order $\lambda(\neq 1)$ (λ is either α or β) are obtained by taking logarithm of λ-order entropic moment [115, 58]. In spherical polar coordinate these can be written in following simplified form by some straightforward mathematical manipulation [97],

$$\begin{aligned}
R_{\mathbf{r}}^{\lambda} &= \frac{1}{1-\lambda} \ln \left(\int_{\mathcal{R}^3} \rho^{\lambda}(\mathbf{r}) \mathrm{d}\mathbf{r} \right) \\
&= \frac{1}{(1-\lambda)} \ln \left(2\pi \int_0^{\infty} [\rho(r)]^{\lambda} r^2 \mathrm{d}r \int_0^{\pi} [\chi(\theta)]^{\lambda} \sin\theta \mathrm{d}\theta \right) \\
&= \frac{1}{(1-\lambda)} \left(\ln 2\pi + \ln[\omega_r^{\lambda}] + \ln[\omega_{(\theta,\phi)}^{\lambda}] \right), \\
R_{\mathbf{p}}^{\lambda} &= \frac{1}{1-\lambda} \ln \left[\int_{\mathcal{R}^3} \Pi^{\lambda}(\mathbf{p}) \mathrm{d}\mathbf{p} \right] \\
&= \frac{1}{(1-\lambda)} \ln \left(2\pi \int_0^{\infty} [\Pi(p)]^{\lambda} p^2 \mathrm{d}p \int_0^{\pi} [\chi(\theta)]^{\lambda} \sin\theta \mathrm{d}\theta \right) \\
&= \frac{1}{(1-\lambda)} \left(\ln 2\pi + \ln[\omega_p^{\lambda}] + \ln[\omega_{(\theta,\phi)}^{\lambda}] \right).
\end{aligned} \tag{58}$$

Here ω_τ^λs are entropic moments in τ (r or p or θ) space with order λ, having forms,

$$\omega_r^\lambda = \int_0^\infty [\rho(r)]^\lambda r^2 \mathrm{d}r, \quad \omega_p^\lambda = \int_0^\infty [\Pi(p)]^\lambda p^2 \mathrm{d}p, \quad \omega_{(\theta,\phi)}^\lambda = \int_0^\pi [\chi(\theta)]^\lambda \sin\theta \mathrm{d}\theta. \tag{59}$$

3.3. Fisher Information (I)

The idea of entropy can adequately explain the degree of disorder of a given phenomenon. Apart from that, however, it is necessary to find out a suitable measure of disorder whose variation derives the event. The concept of entropy is not able to do this. However, I can serve as a good candidate in this context, having the ability to estimate a parameter. Hence it becomes a cornerstone of the statistical field of study, called parameter estimation [126]. I measures the expected error in a smart measurement. Let e^2 be the mean-square error in an estimation of $\hat{\theta}$ then it obeys the relation,

$$e^2 I \geq 1; \quad I \geq \frac{1}{e^2}. \tag{60}$$

Equation (60) suggests that, I is always greater than the reciprocal of e^2; only in case of *Gaussian* distribution it becomes equal to inverse of e^2. The general form of I is,

$$I = \int \frac{|\nabla \rho(\tau)|^2}{\rho(\tau)} \mathrm{d}\tau, \tag{61}$$

which is a gradient functional of density, measuring the local density fluctuation in a given space. In case of a sharp distribution I is higher; whereas, for a flat distribution it is lower. Thus, it can be concluded that, with a rise in uncertainty, I decreases. It resembles the Weizsäcker kinetic energy functional, $T_\omega[\rho]$ often used in DFT [127, 104].

In case of a central potential $I_{\mathbf{r}}$ and $I_{\mathbf{p}}$, the net Fisher information, in $\mathbf{r}$ and $\mathbf{p}$ spaces respectively, are expressed as [128],

$$
\begin{aligned}
I_{\mathbf{r}} &= \int_{\mathcal{R}^3} \left[\frac{|\nabla\rho(\mathbf{r})|^2}{\rho(\mathbf{r})}\right] \mathrm{d}\mathbf{r} = 4\langle p^2\rangle - 2(2l+1)|m|\langle r^{-2}\rangle \\
I_{\mathbf{p}} &= \int_{\mathcal{R}^3} \left[\frac{|\nabla\Pi(\mathbf{p})|^2}{\Pi(\mathbf{p})}\right] \mathrm{d}\mathbf{p} = 4\langle r^2\rangle - 2(2l+1)|m|\langle p^{-2}\rangle, \\
\langle r^2\rangle &= \int_0^{r_c} \psi^*_{n_r,l}(r) r^4\, \psi_{n_r,l}(r)\mathrm{d}r, \quad \langle p^2\rangle = \int_0^{r_c} \psi^*_{n_r,l}(r)[-\nabla^2\psi_{n_r,l}(r)]r^2\mathrm{d}r \\
\left\langle \frac{1}{r^2}\right\rangle &= \int_0^{r_c} \psi^*_{n_r,l}(r)\psi_{n_r,l}(r)\mathrm{d}r, \quad \left\langle \frac{1}{p^2}\right\rangle = \int_0^{r_c} \psi^*_{n_r,l}(p)\psi_{n_r,l}(p)\mathrm{d}p.
\end{aligned}
\tag{62}
$$

The above equations can be further recast in following equivalent forms [129, 130],

$$
\begin{aligned}
I_{\mathbf{r}} &= 8\mathcal{E}_{n,l} - 8\langle v(r)\rangle - 2(2l+1)|m|\langle r^{-2}\rangle \\
I_{\mathbf{p}} &= 8\mathcal{E}_{n,l} - 8\langle T\rangle - 2(2l+1)|m|\langle p^{-2}\rangle \\
&= \frac{8}{\omega^2}\langle v(r)\rangle - 2(2l+1)|m|\langle p^{-2}\rangle,
\end{aligned}
\tag{63}
$$

where $v(p)$ is the p-space counterpart of $v(r)$.

In case of a 3DCHO, I's in r and p space can be expressed analytically as [131],

$$
I_{\mathbf{r}}(\omega) = \frac{\omega}{\sqrt{2}} I_{\mathbf{r}}(\omega = 1), \qquad I_{\mathbf{p}}(\omega) = \frac{\sqrt{2}}{\omega} I_{\mathbf{p}}(\omega = 1). \tag{64}
$$

Thus, an increase in ω leads to rise in $I_{\mathbf{r}}(\omega)$ and fall in $I_{\mathbf{p}}(\omega)$. However, it is obvious that I_t $(= I_{\mathbf{r}} I_{\mathbf{p}})$ remains invariant with ω. Throughout the article, for brevity, $I_{\mathbf{r}}(\omega = 1)$ and $I_{\mathbf{p}}(\omega = 1)$ will be symbolized as $I_{\mathbf{r}}$, $I_{\mathbf{p}}$ respectively.

When $m = 0$, $I_{\mathbf{r}}$ and $I_{\mathbf{p}}$ in Eq. (62) reduce to further simplified forms as below,

$$
I_{\mathbf{r}} = 4\langle p^2\rangle, \qquad I_{\mathbf{p}} = 4\langle r^2\rangle. \tag{65}
$$

It is seen that, at fixed n_r, l, both $I_{\mathbf{r}}$, $I_{\mathbf{p}}$ provide maximum values when $m = 0$, and both of them decrease with rise in m. Hence one obtains the following upper bound for I_t,

$$
I_{\mathbf{r}} I_{\mathbf{p}} \; (= I_t) \leq 16\langle r^2\rangle\langle p^2\rangle. \tag{66}
$$

Further adjustment using Eq. (65) leads to following uncertainty relations [128],

$$
\frac{81}{\langle r^2\rangle\langle p^2\rangle} \leq I_{\mathbf{r}} I_{\mathbf{p}} \leq 16\langle r^2\rangle\langle p^2\rangle. \tag{67}
$$

Therefore, in a central potential, I-based uncertainty product is bounded by both upper as well as lower limits. They are state-dependent, varying with n_r and l quantum numbers.

3.4. Onicescu Energy (E)

Information energy is the counterpart of information entropy. In 1966, Onicescu defined this quantity as [133],

$$E = \sum_i p_i^2. \tag{68}$$

It is interesting to note that, in case of a certain event E is maximum. On the contrary, for an equiprobable distribution it is minimum [134, 127, 135]. Qualitatively it is inverse of S. When S increases, E decreases and vise-versa. In case of continuous probability distribution, E is an expectation value of probability density. Its mathematical form is,

$$E = \int \rho^2(\tau)\, \mathrm{d}\tau. \tag{69}$$

In addition, *Disequilibrium* (D_e) for a continuous probability distribution is defined as,

$$De = \int_{-\infty}^{+\infty} \rho^2(\tau)\mathrm{d}\tau. \tag{70}$$

Therefore, in this occasion, disequilibrium and information energy have same form. Like the previous measures, it is also utilized in orbital-free DFT [105], testing normality [136], electron correlation [110], Colin conjecture [100, 101], configuration interaction [137] etc.

By definition, E refers to the 2nd-order entropic moment [127]; for central potential it assumes the form (E_t is the Onicescu energy product),

$$\begin{aligned} E_r &= \int_0^\infty [\rho(r)]^2 r^2 \mathrm{d}r, \;\; E_p = \int_0^\infty [\Pi(p)]^2 p^2 \mathrm{d}p, \\ E_{\theta,\phi} &= \int_0^\pi [\chi(\theta)]^2 \sin\theta \mathrm{d}\theta. \;\; E_t = E_r E_p E_{\theta,\phi}^2. \end{aligned} \tag{71}$$

Uncertainty product for such measures are studied in [138].

3.5. Complexities

In a system, statistical complexity arises due to breakdown of symmetry. It illustrates a competing effect of two complementary quantities, offering a qualitative idea of the organization, structure and correlation. According to [139] it can be broadly characterized into three categories: (i) advances monotonically with disorder (ii) reaches its minimal value for both completely ordered and disordered systems, and a maximum at some intermediate level (iii) increases with order. It has finite value in a state lying between two limiting cases of complete order (maximum distance from equilibrium) and maximum disorder (at equilibrium). Figure 1 pictorially demonstrates above three different kind of complexities as a function of disorder.

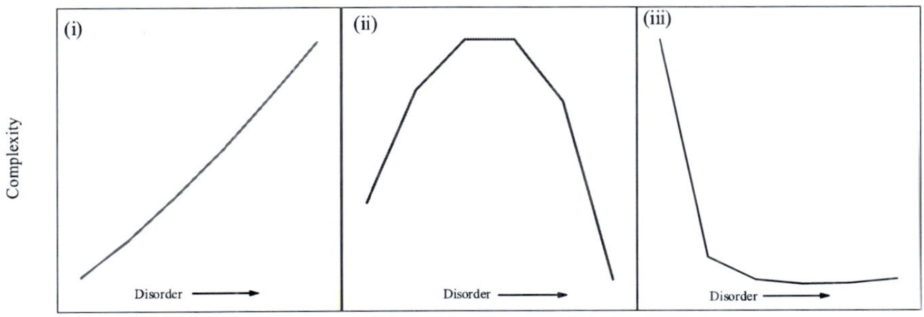

Figure 1. Three different categories of Complexities, as functions of disorder.

The statistical measure of complexity (C_{LMC}) is nothing but the product of the information content (H) (such as S, R etc.) and concentration of spatial distribution (D), and can be written as $C_{LMC} = H.D$. This was later criticized [140] and modified [141] to the form of $C_{LMC} = D.e^S$, in order to satisfy few conditions such as reaching minimal values for both extremely ordered and disordered limits, invariance under scaling, translation and replication. Various definitions were put forth in literature. Some notable ones include Shiner, Davidson and Landsberg (SDL) [142, 139], Fisher-Shannon (C_{IS}) [143, 144, 145], Cramér-Rao [145, 146], generalized Rényi-like [147, 148, 149] complexity, etc. Amongst these, C_{IS} corresponds to a measure which probes a system in terms of complementary global and local factors, and also satisfies certain desirable properties in complexity [127], like invariance under translations and re-scaling transformations, invariance under replication, near-continuity, etc. This has remarkable applications in the study of atomic shell structure, ionization processes [144, 145, 150], as well as in molecular properties like energy, ionization potential, hardness, dipole moment in the localization-delocalization plane showing chemically significant pattern [151], molecular reactivity studies [152]. Some elementary chemical reactions such as hydrogenic-abstraction reaction [153], identity SN^2 exchange reaction [154], and also concurrent phenomena occurring at the transition region [155] of these reactions have been investigated through composite information-theoretic measures in conjugate spaces.

Without any loss of generality, let us define complexity in following general form $C = Ae^{b.B}$. The order (A) and disorder parameters (B) may include (E, I) and (R, S) respectively. With this in mind, we are interested in the following four quantities,

$$C_{ER} = Ee^{bR}, \qquad C_{IR} = Ie^{bR}, \qquad C_{ES} = Ee^{bS}, \qquad C_{IS} = Ie^{bS}. \quad (72)$$

3.6. Relative Information

Kullback-Leibler divergence or relative entropy is a descriptor or quantifier of how a measured probability distribution function deviates from a given reference distribution [156, 157]. In quantum mechanics, this characterizes a measure of distinguishability between two states. It actually quantifies the change of information from one state to other [126]. Relative R and S was explored for various atomic systems using H atom ground state as reference [158]. A detailed investigation reveals that, they are directly related with atomic radii and quantum capacitance [159, 158]. Another interesting measure is the relative Fisher information (IR) [160]. It has importance in different topics of physics and chemistry, such as to calculate phase-space gradient of dissipated work and information [161], deriving Jensen divergence [162], relation with score function [163], in the context of study of probability current [164], in thermodynamics [165], etc. Of late it has been successfully used in formulating atomic densities [166] and deriving density functionals under local-density and generalized-gradient approximations [167]. Further, IR along with Hellmann-Feynman and virial theorem has been used to develop a Legendre transform structure related to SE [168]. It has been designed self-consistently on the basis of estimation theory [169]. In quantum chemistry perspective, it has been successfully formulated using above two theorems and entropy maximization principle [170, 171, 172]. Very recently, IR for some exactly solvable potentials including 1D and 3D QHO in both position and momentum spaces has been estimated analytically using ground state of definite symmetry (for example, $l = 0$ for s orbitals, $l = 1$ for p orbital, and so on) as reference [173]. Before that, a numerical estimation of IR in position space is done for H-atom using $1s$ orbital as basis [174].

For two normalized probability densities $\rho_{n,l,m}(\tau)$, $\rho_{n_1,l_1,m_1}(\tau)$, IR is expressed as,

$$\text{IR}\left[\rho_{n,l,m}(\tau)|\rho_{n_1,l_1,m_1}(\tau)\right] = \int_{\mathcal{R}^3} \rho_{n,l,m}(\tau) \left| \nabla \ln \left\{ \frac{\rho_{n,l,m}(\tau)}{\rho_{n_1,l_1,m_1}(\tau)} \right\} \right|^2 \mathrm{d}\tau. \quad (73)$$

Here n, l, m and n_1, l_1, m_1 are the descriptors of target and reference states respectively, while τ is a generalized variable. In case of central potential, these probability densities $\rho_{n,l,m}(\tau)$, $\rho_{n_1,l_1,m_1}(\tau)$ can be expressed in following forms, without any loss of generality,

$$\rho_{n,l,m}(\tau) = R^2_{n,l}(s)\Theta^2_{l,m}(\theta); \quad \rho_{n_1,l_1,m_1}(\tau) = R^2_{n_1,l_1}(s)\Theta^2_{l_1,m_1}(\theta). \quad (74)$$

In the above equation, $R_{n,l}(s)$, $R_{n_1,l_1}(s)$ signify radial parts, $\Theta_{n,l}(\theta)$, $\Theta_{n_1,l_1}(\theta)$ represent angular contributions of two wave functions, whereas "s" implies either r or p variable in respective radial functions. Thus Eq. (73) may be rewritten as,

$$\begin{aligned}\text{IR}\left[\rho_{n,l,m}(\tau)|\rho_{n_1,l_1,m_1}(\tau)\right] &= \text{IR}\left[\rho_{n,l}(s)|\rho_{n_1,l_1}(s)\right] + \left\langle \frac{1}{s^2} \right\rangle \text{IR}\left[\Theta^2_{l,m}(\theta)|\Theta^2_{l_1,m_1}(\theta)\right] + \\ 2\int_0^\infty s\, R^2_{n,l}(s) \left[\frac{d}{ds} \ln \left\{ \frac{R^2_{n,l}(s)}{R^2_{n_1,l_1}(s)} \right\}\right] \mathrm{d}s &\int_0^\pi \Theta^2_{l,m}(\theta) \left[\frac{d}{d\theta} \ln \left\{ \frac{\Theta^2_{l,m}(\theta)}{\Theta^2_{l_1,m_1}(\theta)} \right\}\right] \sin\theta \mathrm{d}\theta\end{aligned} \quad (75)$$

where the following quantities have been defined,

$$\begin{aligned}\text{IR}\left[\rho_{n,l}(s)|\rho_{n_1,l_1}(s)\right] &= \int_0^\infty s^2\, \rho_{n,l}(s) \left| \frac{d}{ds} \ln \left\{ \frac{\rho_{n,l}(s)}{\rho_{n_1,l_1}(s)} \right\} \right|^2 \mathrm{d}s \\ \text{IR}\left[\Theta^2_{l,m}(\theta)|\Theta^2_{l_1,m_1}(\theta)\right] &= \int_0^\pi \Theta^2_{l,m}(\theta) \left| \frac{d}{d\theta} \ln \left\{ \frac{\Theta^2_{l,m}(\theta)}{\Theta^2_{l_1,m_1}(\theta)} \right\} \right|^2 \sin\theta\ \mathrm{d}\theta\end{aligned} \quad (76)$$

4. Result and Discussion

4.1. 1D Confined Harmonic Oscillator (1DCHO)

Panel (a) of Figure 2 represents an SCHO scenario for two box lengths 1 and 2, respectively. Next, the asymmetric confinement in a 1DQHO can be invoked in two different ways: (i) by varying the box boundary, keeping box length and d_m fixed at zero (ii) the other way is to change d_m by keeping box length and boundary constant. We have chosen the second option; where a rise in d_m transfers the minimum towards the right of origin, keeping box length, $b_2 - b_1 = 2$ fixed, while left and right boundaries are residing at $b_1 = -1$ and $b_2 = 1$. Moreover, since $\frac{1}{2}k(x \pm d_m)^2$ illustrates a mirror-image pair potential trapped within b_1 and b_2, their eigenvalues and expectation values are same for all states. As the respective wave functions are mirror images of each other, it suffices to consider the behavior of either one of them. The other one automatically follows from it. The right-hand panel (b) of Figure 1 shows a schematic representation of an ACHO potential, at five different d_m values.

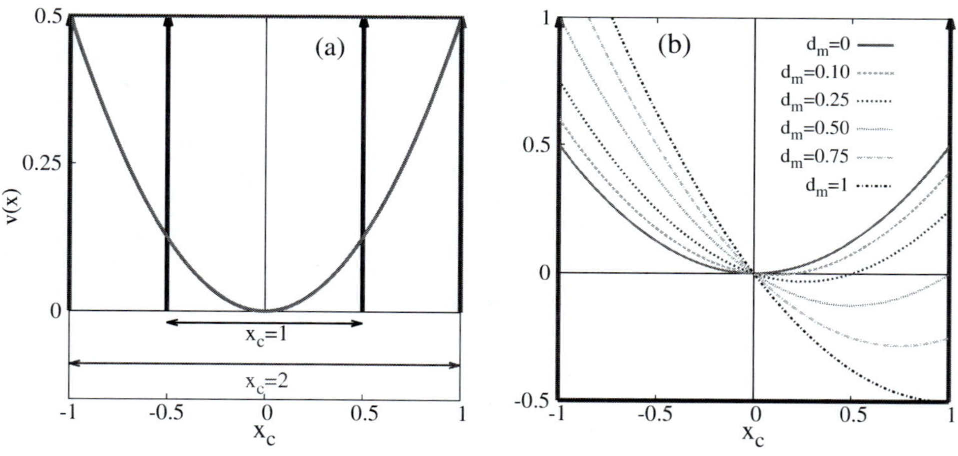

Figure 2. Schematic representation of a confined 1DQHO potential: (a) SCHO at two separate box lengths (b) ACHO at six chosen values of d_m [132].

4.1.1. Symmetrically Confined Harmonic Oscillator (SCHO)

In this case, all calculations are performed involving the wave function given in Eq. (2). Figure 3 shows plots of S_x, S_p versus x_c, for first five states of SCHO.

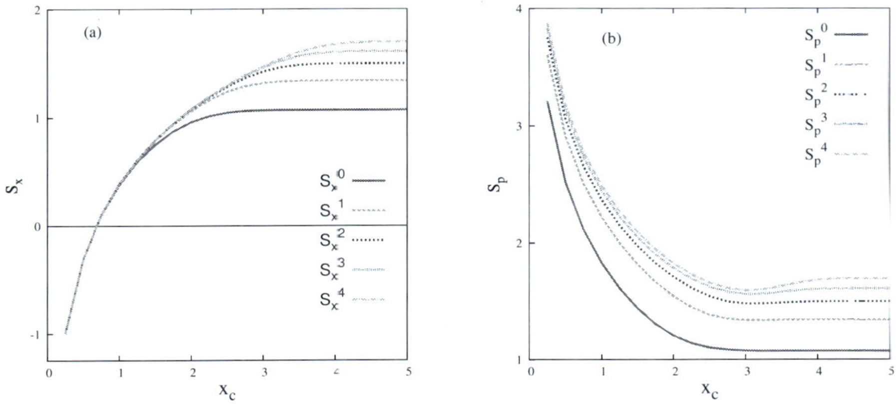

Figure 3. Plot of (a) S_x (b) S_p, of SCHO potential, as function of x_c, for first five states [132].

In panel (a), S_x progresses with box length and converges to a constant value (S_x of 1DQHO) at sufficiently large box length. At small x_c region, S_x changes rather insignificantly with n, corresponding to the behavior of a PIB problem (where S_x remains unchanged with n). Next, panel (b), imprints that S_p abates with increase in box length and finally merges to corresponding 1DQHO value. It is important to note that, there appears a minimum in S_p for all excited states. Appearance of such minimum may be ascribed to the competing effect in p space. In fact, three possibilities could be contemplated (l_x, l_p are box lengths in x, p space):

(a) When $l_x \to 0$, then $l_p \to \infty$.

(b) When l_x is finite, then l_p is also finite. But an increase in l_x leads to a decrease in l_p.

(c) When $l_x \to \infty$, then also $l_p \to \infty$.

These plots suggest that, at the beginning, with increase in l_x, particle gets localized in p space (S_p decreases), but when potential behaves like 1DQHO, de-localization predominates. Thus, existence of minimum in S_p is due to presence of two conjugate forces. These variations of S_x, S_p with x_c are in complete agreement with the findings of [175].

The behavior of I_x, I_p with respect to x_c of first five stationary states are exhibited in two bottom panels (a), (b) of Figure (4). One can see that initially

I_x decreases sharply, extent of which is maximum for the lowest state and successively decreases with n. Then it attains a constant value after a certain x_c. On the contrary, in panel (b), I_p strongly grows with x_c at first; the extent increases with n, lowest state producing lowest. Finally, for all states, it eventually reaches a state-dependent constant value as in I_x. It is pertinent to mention that, behavior of I_x, I_p under confinement agrees with those of Δp, Δx [175]. Next, panel (c) reveals that, E_x for all these states remain very close to each other at smaller x_c; then as x_c progresses, E_x for individual states branch out and decreases indicating delocalization. In the end, it reaches some constant value for all n. Lastly, in panel (d), E_ps tend to advance in the beginning for all states at low x_c region, eventually becoming smooth after some threshold x_c. These calculation have further consolidated the conclusions obtained from the study of S_x, S_p, S demonstrated earlier.

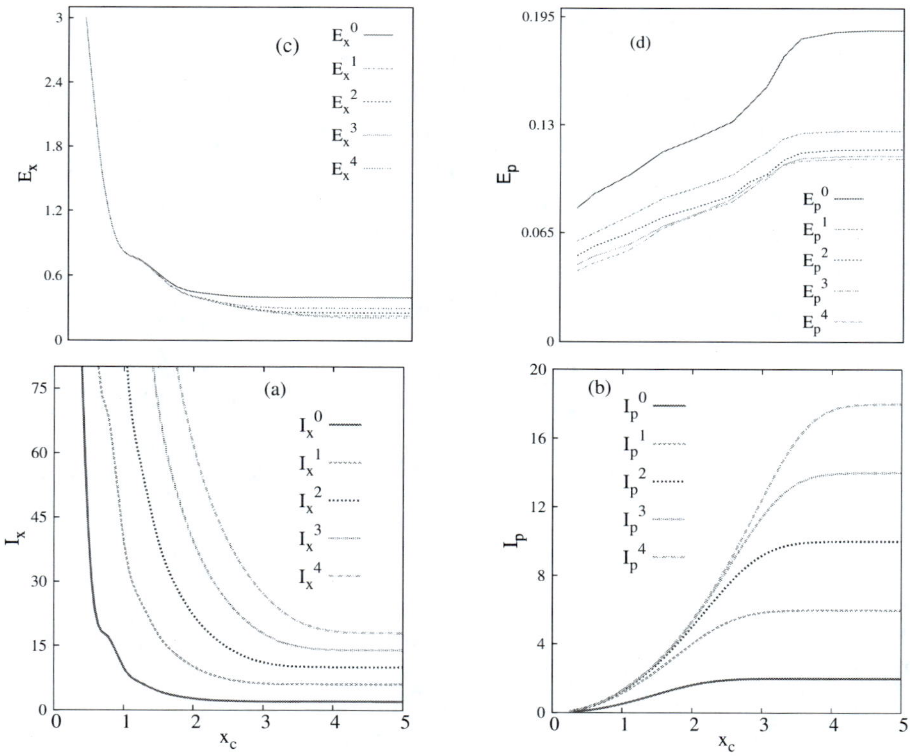

Figure 4. Plot of I_x, I_p, E_x, E_p for first five states of SCHO potential, as functions of x_c, in panels (a), (b), (c) and (d). See text for details [132].

Before looking into the behavior of IE in ACHO, it is pertinent to make some comments about its eigenvalues and eigenfunctions. ACHO has been studied using power-series solution [32] and ITP method [34]. In this section, we shall, however, present another simple method, which produces quite accurate results. In this *variation induced exact diagonalization* procedure [176], an energy functional is minimized using an SCHO basis set. Recently this method has been successfully employed in studying symmetric and asymmetric double-well potentials [177, 178], where a 1DQHO basis was utilized.

The Hamiltonian matrix elements are evaluated by using functions given in Eq. (2). Presence of a single non-linear parameter α allows us to adopt a coupled variation procedure. Further it is easy to confirm the convergence of results with respect to basis dimension N. Thus it provides a secular equation at each α. The kinetic energy part blows up to ∞ when $\alpha \rightarrow \infty$, whereas at $\alpha \rightarrow 0$, potential energy part behaves in a similar fashion. This qualitative analysis through uncertainty principle, confirms the existence of such basis. Diagonalization of $H_{mn} = \langle m|\hat{H}|n\rangle$ leads to accurate eigenvalues and eigenfunctions, which is attained through MATHEMATICA package. In principle, to achieve exact solution, one needs to employ the *complete* basis; however for practical purposes a truncated basis of finite dimension is envisaged. Here, $N = 50$ appears adequate; with further increase in basis, result improves. A cross-section of energies obtained from above scheme is produced in Table 3, for lowest six states of ACHO at four selected d_m. Some literature results are provided for comparison, wherever available. For all four d_m, present energies with SCHO basis, practically coincide with ITP results [34], for all digits reported. It is expected that this SCHO basis may be useful in future for other confined and free 1D potentials.

Our calculated ACHO energies of Table 3 are presented in Figure 5, for first six states with respect to d_m, in panels (a)-(f). Ranges of ϵ_n and d_m are different for each n. Variation in ϵ_0 is unique amongst all ϵ_n, where, from an initial higher value at small d_m, ground-state energy continuously falls. In all excited states, however, it passes through a maximum; with increase in n, this maximum shifts to higher values of d_m. Next Figure 6 portrays our computed wave functions for $n = 0-5$ states of ACHO at four selected d_m values. Clearly, the maximum, minimum and nodal positions of all ψ_n's switch towards right with advancement in d_m. These plots suggest that particle gets localized in x space as d_m progresses.

Table 3. Comparison of energies of first six states of ACHO potential at four distinct d_m. Length of the box is fixed at 2. PR implies Present Result [132]. See text for details

d_m	ϵ_0 (PR)	ϵ_0 (Ref.)	ϵ_1 (PR)	ϵ_1(Ref.)
0.36	2.7177633960054	2.7177633960054‡	10.283146010610	10.283146010610‡
1.92	6.0383021056781	6.0383021056781‡	13.901445986629	13.901445986629‡
5.00	26.065225076406	26.065225076406‡	35.462261039378	35.462261039378‡
10.0	97.474035270680	97.474035270680‡	110.51944554927	110.51944554927‡
	ϵ_2 (PR)	ϵ_2 (Ref.)	ϵ_3 (PR)	ϵ_3(Ref.)
0.36	22.648848755052	22.648848755052‡	39.929984298830	39.929984298830‡
1.92	26.249310409373	26.249310409373‡	43.513981920357	43.513981920357‡
5.00	47.817024422796	47.817024422796‡	64.900200447511	64.900200447511‡
10.0	123.593144939095	123.593144939095‡	140.555432078323	140.555432078323‡
	ϵ_4 (PR)	ϵ_4 (Ref.)	ϵ_5 (PR)	ϵ_5(Ref.)
0.36	62.140768627508	62.140768627508‡	89.284409553063	89.284409553063‡
1.92	65.715672311936	65.715672311936‡	92.854029622882	92.854029622882‡
5.00	87.137790461503	87.137790461503‡	114.244486402564	114.244486402564‡
10.0	162.519960161732	162.519960161732‡	189.515389275133	189.515389275133‡

‡ITP result [34].

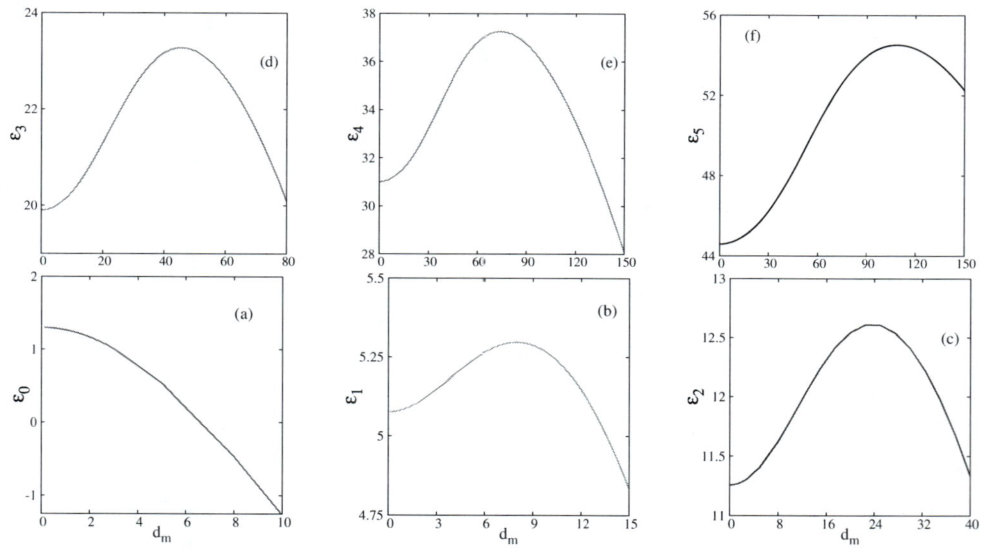

Figure 5. Variation of ϵ_n $(n=0-5)$ of ACHO potential as function of d_m [132].

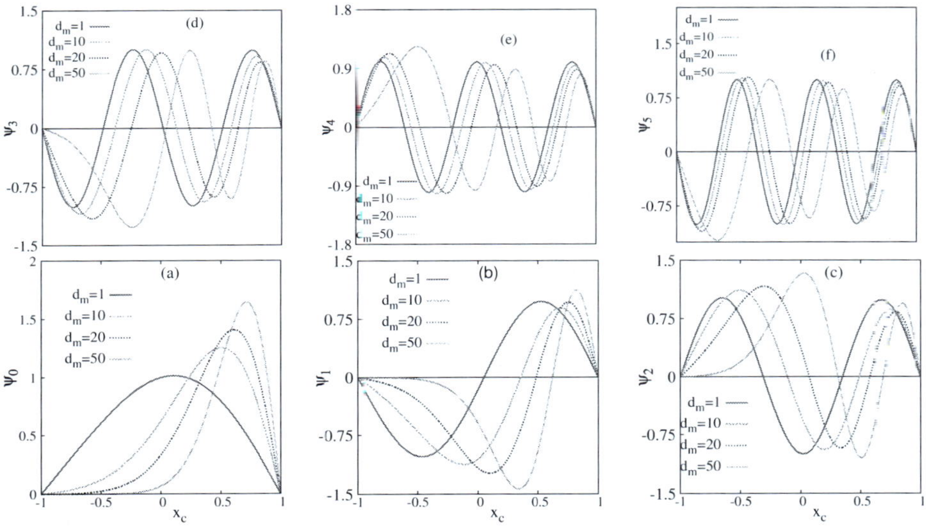

Figure 6. Wave functions of lowest six states of ACHO potential at four specific values of d_m, namely 1, 10, 20, 50. Panels (a)–(f) represent $n = 0 - 5$ states [132]. See text for details.

4.1.2. Asymmetrically Confined Harmonic Oscillator (ACHO)

Now, the focus is on the ACHO case. This presentation is based on results of ϵ_n; S_x, S_p; I_x, I_p; E_x, E_p as functions of d_m, for some low-lying states. Throughout the whole presentation, the box length has been kept fixed at 2 and boundaries are placed at -1 and 1. At the outset it is important to note that, an increase in d_m leads to localization in real space. Thus it can be expected that, there will be delocalization in p space.

Now we move on to I_x, I_p variations with change of d_m. These are shown pictorially in Figure 7 left (a), and right (b) panels display these in conjugate x, p spaces respectively. For ground and first excited state I_x tends to increase with d_m, on the whole, whereas for ($n = 2$) state, the same decreases. Thus, for first two states, it indicates localization in x space and delocalization for 2nd excited state. A careful investigation of panels b(i)–b(iii) of Figure 7 reveals that, I_p, for all n under consideration, consistently decreases with increase of d_m, signifying a delocalization of particle in p space.

Next, it is imperative to explore changes in behavior of S_x, S_p with respect to d_m. These are offered in Figure 8. It is clear from left panels a(i)–a(iii) of Fig-

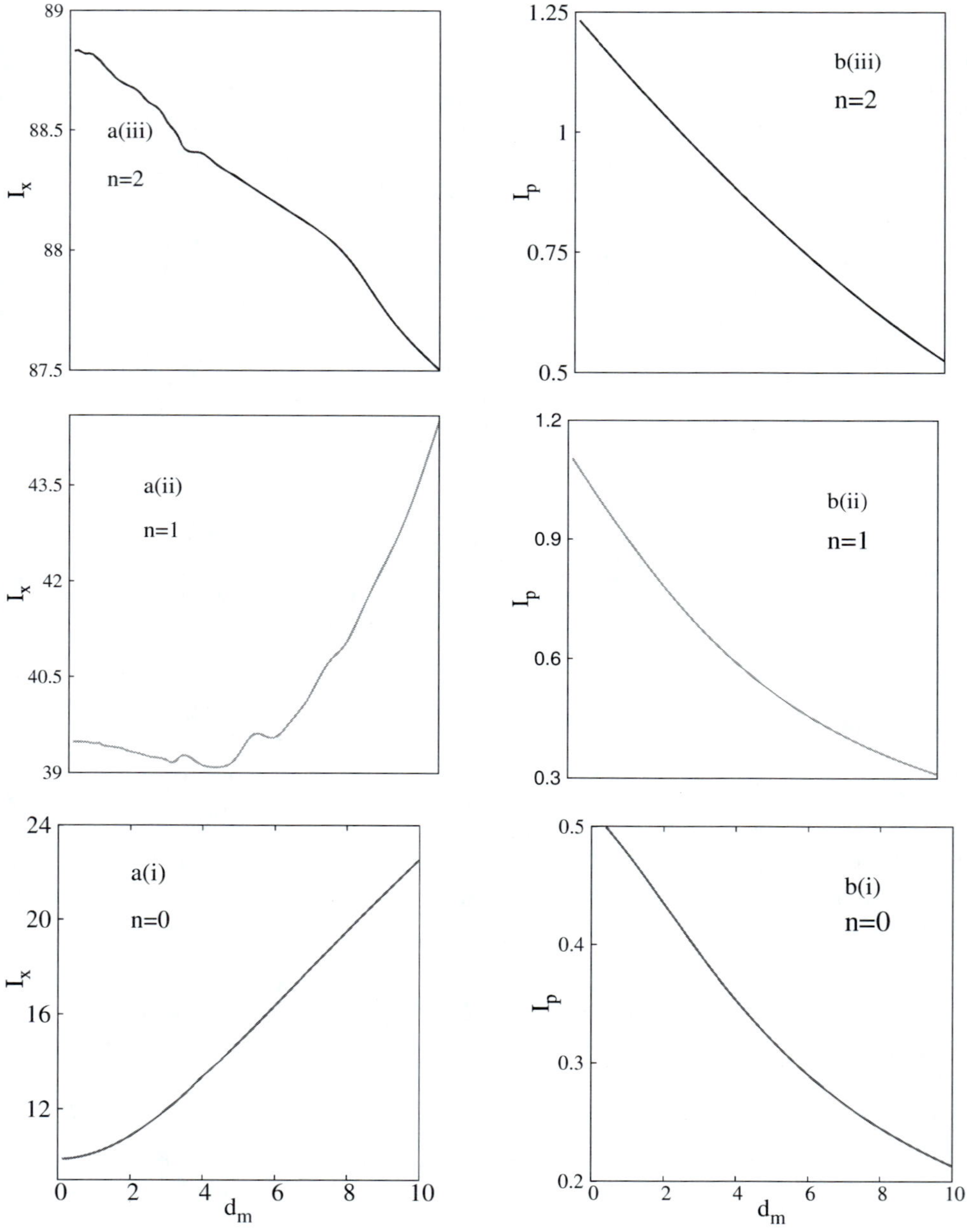

Figure 7. Plot of I_x, I_p, I of first five states of ACHO potential, as function of d_m, in left (a), middle (b), right (c) columns; (i)–(iii) represent $n = 0 - 2$ states [132]. See text for details.

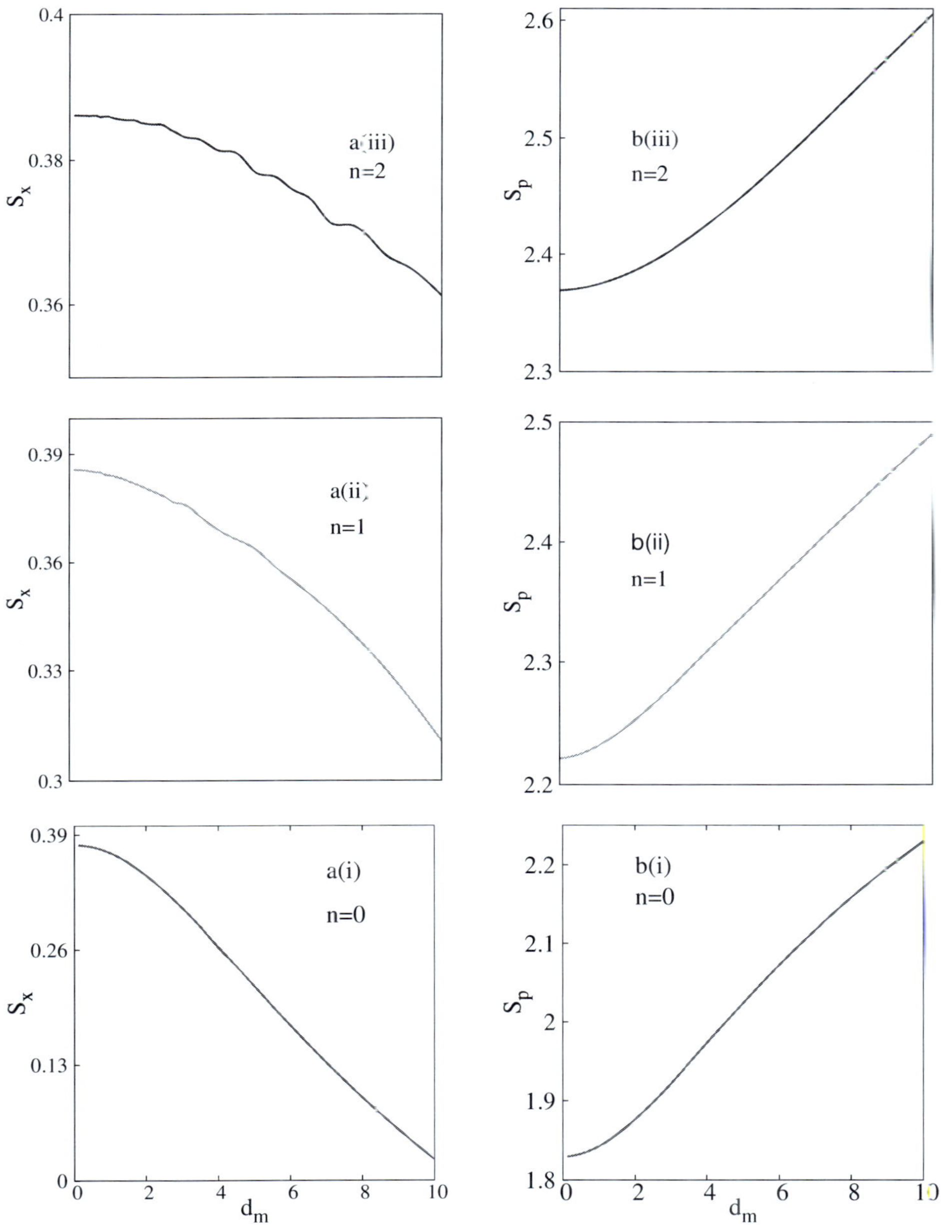

Figure 8. Plot of S_x, S_p, S of first three states of ACHO potential, as function of d_m, in left (a), middle (b), right (c) columns; (i)–(iii) represent $n = 0 - 2$ states [132]. See text for details.

ure 8 that, S_x for all these three states generally decreases monotonically with growth in d_m, signifying localization of particle in position space. Likewise, a scrutiny of right plots b(i)–b(iii) of Figure 8 summarizes an opposite trend in S_p with increase in d_m. A sample of S_x, S_p are reported in Table 4 for $n = 0, 1, 2$ states at five selected d_m values, namely, 0.12, 2.04, 5, 7 and 10. Evidently, these entries corroborate the outcomes of Figure 8.

Now, we move on to an analysis of E, using Figure 9, for $n = 0 - 2$. Panels a(i)–a(iii) suggest that, for all three states there is an overall increase of E_x with rise in d_m. This indicates localization in x space. Panels b(i)–b(iii) portray that, general trend of E_p is a gradual decrease with rise in d_m.

On the basis of above discussion, it is clear that, only S_x, S_p can adequately explain localization-delocalization phenomena in an ACHO. Study of E_x, E_p can also offer valuable knowledge about the dual nature of E in composite x, p space. But, I_x, I_p appear to be inadequate in explaining the contrasting phenomena in ACHO.

4.2. 3D Confined Harmonic Oscillator (3DCHO)

At the onset, it is convenient to point out that, *net* information measures in conjugate r and p space may be segmented into radial and angular parts. In a given space, the results provided correspond to *net* measures including the *angular* contributions. One can transform the 3DQHO into a 3DCHO by squeezing the radial boundary of former from infinity to a finite region. This alteration in radial environment does not affect the *angular* boundary conditions. Therefore, angular part of the information measures in free and confined systems remain unchanged in both spaces. Further as we are solely focused in *radial* confinement, this will also not influence the characteristics of a given measure as one changes r_c. Throughout this investigation, magnetic quantum number m remains fixed to 0 for R, S, E calculation. However, I has been studied for *non-zero* m states. The radial wave function in r and p spaces depend only on n_r, l quantum numbers. Hence, in both space, radial wave function can be obtained by considering $m = 0$. Further, a change in m from zero to non-zero value will not influence the expression of radial wave function in p space. Also note that, we have followed the spectroscopic notation, i.e., the levels are denoted by $n_r + 1$ and l values (see, e.g., [69]). Therefore, $n_r = 0$ and $l = 4$ signifies $1g$ state. The radial quantum number n_r relates to n as $n = 2n_r + l$.

$I_{\mathbf{r}}$, $I_{\mathbf{p}}$ values are obtained from Eq. (62). In all occasions, it has been ver-

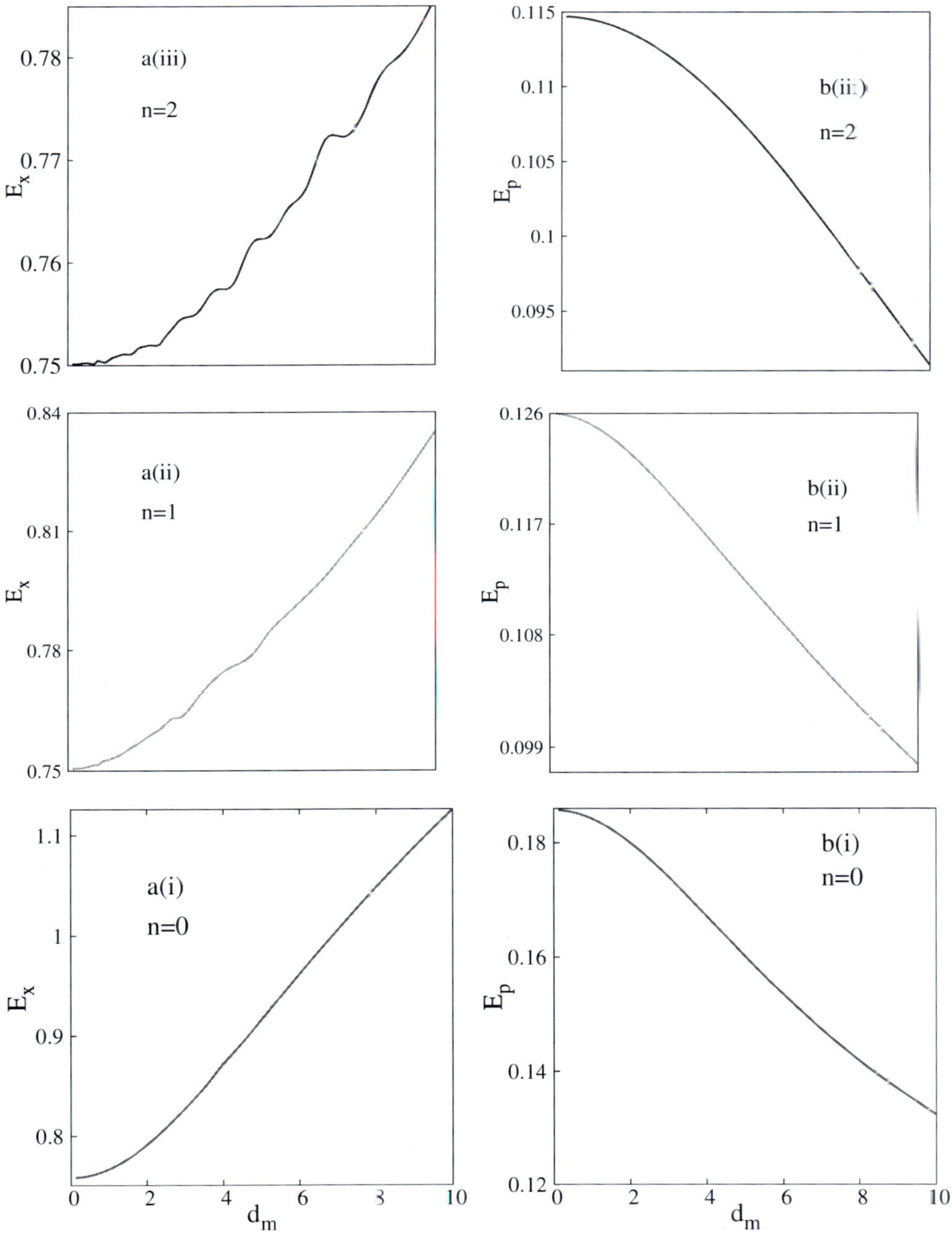

Figure 9. Plot of E_x, E_p, E of first three states of ACHO potential, as function of d_m, in left (a), middle (b), right (c) columns; (i)–(iii) represent $n = 0 - 2$ states [132]. See text for details.

Table 4. S_x, S_p, S **for** $n{=}0-2$ **states of ACHO at five specific** d_m **[132]. See text for details**

d_m	S_x^0	S_p^0	S^0	S_x^1	S_p^1	S^1	S_x^2	S_p^2	S^2
0.12	0.3783	1.8296	2.2079	0.3857	2.2212	2.6069	0.3862	2.3692	2.7554
2.04	0.3428	1.8779	2.2208	0.3807	2.2512	2.6319	0.3850	2.3852	2.7703
5.0	0.2184	2.0238	2.2422	0.3636	2.3390	2.7027	0.3782	2.4512	2.8295
8.0	0.093	2.1576	2.2506	0.3360	2.4313	2.7674	0.3695	2.5429	2.9124
10.0	0.0233	2.2294	2.2527	0.3108	2.4900	2.8009	0.3611	2.6057	2.9668

ified that, in both spaces, as $r_c \rightarrow \infty$, $I_{\mathbf{r}}$ and $I_{\mathbf{p}}$ merge to respective 3DQHO limit. The net I in r and p spaces are divided into radial and angular part. But in both $I_{\mathbf{r}}$ and $I_{\mathbf{p}}$ expressions, angular contribution is normalized to unity. Hence, evaluation of all these targeted quantities using only radial part will serve our purpose. Pilot calculations are done for $1s$-$1g$ and $2s$-$2m$ states, with r_c varying from 0.1 to 7 a.u. The former set is chosen as they represent node-less ground states corresponding to various l, whereas, $2s$-$2m$ states are considered to perceive the effect of nodes on I at non-zero m.

It is worthwhile noting that, I's can be accurately calculated from a knowledge of $\langle p^2\rangle, \left\langle \frac{1}{r^2}\right\rangle, \langle r^2\rangle, \left\langle \frac{1}{p^2}\right\rangle$. Two possibilities may be invoked: (i) first three quantities evaluated in r space, while $\langle p^{-2}\rangle$ in p space (ii) $\langle r^2\rangle$, $\langle r^{-2}\rangle$ in r space, while $\langle p^2\rangle$, $\langle p^{-2}\rangle$ in p space. Here we have opted for first route removing the necessity to do numerical differentiation in either spaces. This has been discussed in detail in a recent article [97].

Exact analytical form of $I_{\mathbf{r}}$ and $I_{\mathbf{p}}$ in isotropic 3DQHO was given in [128],

$$I_{\mathbf{r}}(\omega) = 4\omega\left(2n_r + l - |m| + \frac{3}{2}\right), \quad I_{\mathbf{p}}(\omega) = \frac{4}{\omega}\left(2n_r + l - |m| + \frac{3}{2}\right). \tag{77}$$

Thus, at a certain m, both $I_{\mathbf{r}}(\omega), I_{\mathbf{p}}(\omega)$ increase as n_r and l approach higher values. Similarly, for specific n_r and l, both $I_{\mathbf{r}}(\omega), I_{\mathbf{p}}(\omega)$ regress with growth in $|m|$. Effect of ω on $I_{\mathbf{r}}(\omega), I_{\mathbf{p}}(\omega)$ is quite straightforward. $I_{\mathbf{r}}(\omega)$ advances and $I_{\mathbf{p}}(\omega)$ decreases with rise of ω. By putting $\omega = 1$ in Eq. (77) one easily recovers expressions for $I_{\mathbf{r}}$, $I_{\mathbf{p}}$ in a 3DQHO.

At first we illustrate the behavior of 3DCHO at $r_c \rightarrow 0$. A diligent study reveals that, at small r_c region, 3DCHO has an energy spectrum comparable to

Table 5. $\mathcal{E}_{n_r,l}(n_r = 0)$ of lowest five circular states of 3DCHO, PISB at five r_c, from [34, 130]

l	$r_c = 0.01$	$r_c = 0.05$	$r_c = 0.1$	$r_c = 0.2$	$r_c = 0.5$
			3DCHO		
0	49348.02202373	1973.92123372	493.48163345	123.37570844	19.77453418
1	100953.64280465	4038.14517967	1009.53830088	252.39159906	40.42827649
2	166087.30959293	6643.49293120	1660.87528919	415.22704789	66.48975653
3	244155.96823805	9753.60193720	2441.56211674	610.39965899	97.72324914
4	334771.55964446	13390.86304096	3347.71822121	836.93939916	133.97424633
			PISB		
0	49348.022005446	1973.92088021	493.48022005	123.37005501	19.73920880
1	100953.64278213	4038.14571128	1009.53642782	252.38410695	40.38145711
2	166087.30957134	6643.49238285	1660.87309571	415.21827392	66.43492382
3	244155.96821809	9753.60153136	2441.55968218	610.38992054	97.66238728
4	334771.55962552	13390.86238502	3347.71559625	836.92889906	133.90852385

that of a particle in a spherical box (PISB). This leaning generally holds good for all other states as well. A cross-section of eigenvalues ($1s$-$1g$) of lowest five circular states corresponding to l = 0-4, of 3DCHO and PISB, given in Table 5, at five selected r_c, *viz.*, $0.01, 0.05, 0.1, 0.2, 0.5$, supports this fact. However this observation should not be misinterpreted to conclude that at $r_c \rightarrow 0$, 3DCHO leads to PISB. Because that can happen only when both systems have nearly equal kinetic energy as well as potential energy components. This is not apparent from this table. At this point, it is noteworthy to mention that, like 3DCHO, PISB is also exactly solvable; eigenfunctions are explicable directly in terms of first-order Bessel function, and given as,

$$J_l(Z) = (-1)^l Z^l \left(\frac{1}{Z} \frac{d}{dZ} \right)^l \left(\frac{\sin Z}{Z} \right), \tag{78}$$

where $Z = \sqrt{\mathcal{E}_{n_r,l}}\, r$. At the boundary when $r = r_c$, $Z = Z_{n_r,l}$ and $J_l(Z_{n_r,l}) = 0$. Moreover, at $r = r_c$, the energy of a (n_r, l) state is written as $\mathcal{E}_{n_r,l} = \frac{Z^2_{n_r,l}}{r_c^2}$ [179]. This $J_l(Z_{n_r,l}) = 0$ is a transcendental equation and at a constant n_r, l, this $Z_{n_r,l}$ is evaluated by the help of MATHEMATICA program package. Thus all the PISB energies in lower segment of this table have been calculated following the above procedure.

Table 6. $I_{\mathbf{r}}$, $I_{\mathbf{p}}$ of lowest five circular states of 3DCHO and PISB at five r_c at fixed $m(0)$ [130]

System	l	$r_c = 0.01$	$r_c = 0.05$	$r_c = 0.1$	$r_c = 0.2$	$r_c = 0.5$
				$I_{\mathbf{r}}$		
3DCHO	0	394784.1761898	15791.36986976	3947.84176	986.960440	157.913740
	1	807629.1424372	32305.16943736	8076.29142	2019.072855	323.05170823
	2	1328698.4767434	53147.9434496	13286.984765	3321.7461915	531.4794285
	4	2678172.4771556	107126.9043276	26781.72476	6695.431192	1071.269013
PISB	0	394784.1760435	15791.3670417	3947.8417604	986.9604401	157.9136704
	1	807629.1422570	32305.1656902	8076.2914225	2019.0728556	323.0516569
	2	1328698.4765707	53147.9390628	13286.9847657	3321.7461914	531.4793906
	4	2678172.4770042	107126.8990801	26781.7247700	6695.4311925	1071.2689908
				$I_{\mathbf{p}}$		
3DCHO	0	0.0001130690	0.0028267272	0.011306900	0.0452270673	0.282533301
	1	0.0001498425	0.0037460639	0.014984249	0.0599366038	0.374503742
	2	0.0001754798	0.0043869959	0.017547979	0.0701916254	0.438623720
	4	0.0002100025	0.0052500632	0.021000250	0.0840008284	0.524961467
PISB	0	0.0001130690	0.002826727	0.011306909	0.045227638	0.282672741
	1	0.0001498425	0.003746064	0.014984256	0.059937024	0.374606402
	2	0.0001754798	0.004386996	0.017547984	0.070191936	0.438699601
	4	0.0002100025	0.005250063	0.021000253	0.084001012	0.525006325

The upper portion of Table 6 portrays $I_{\mathbf{r}}$ of 3DCHO and PISB at five selected r_c values introduced before; the respective $I_{\mathbf{p}}$ of two systems are reported in lower portion. It is interesting to point out that, for 3DQHO and 3DCHO in a state having $m = 0$, $I_{\mathbf{r}}$ and $I_{\mathbf{p}}$ are directly connected to expectation values of kinetic and potential energy as;

$$I_{\mathbf{r}}(\omega) = 8\langle T \rangle, \quad I_{\mathbf{p}}(\omega) = \frac{8}{\omega^2}\langle v(r) \rangle. \tag{79}$$

But for a PISB, total energy is exclusively kinetic energy as the potential energy is zero. However, one can evaluate $I_{\mathbf{r}}$, $I_{\mathbf{p}}$ for PISB and collate the results with 3DCHO. Because, a pair of systems possessing same $I_{\mathbf{r}}$, $I_{\mathbf{p}}$ for all states indicate identical physical and chemical environment. Hence, here we have exploited I to investigate the characteristics of PISB and 3DCHO at $r_c \rightarrow 0$. The

Table 7. $I_{\mathbf{r}}, I_{\mathbf{p}}$ for lowest five n_r ($l = 0$) values at six different r_c [130]. See text for details

n_r	$r_c = 0.1$	$r_c = 0.5$	$r_c = 1$	$r_c = 2$	$r_c = 7$	$r_c = \infty$
			$I_{\mathbf{r}}$			
0	3947.84176	157.9137401	39.48285935	10.130828577	6.00000000	6
1	15791.3670	631.654662	157.91245186	39.42241043	14.00000000	14
2	35530.57584	1421.2230213	355.3049651	88.77709457	22.00000000	22
3	63165.46816	2526.6187189	631.6541846	157.88183967	30.00000000	30
4	98696.0440	3947.8417551	986.960106	246.718681361	38.00000000	38
			$I_{\mathbf{p}}$			
0	0.0113069007	0.2825333012	1.1217967676	3.9877029335	6.0000000	6
1	0.01282672989	0.32070687721	1.28512015257	5.25470219890	14.0000000	14
2	0.0131081767	0.3277292039	1.3124046596	5.34276239849	22.00000000	22
3	0.0132066828	0.3301825760	1.3216621568	5.3463257844	30.00000000	30
4	0.0132522770	0.33131731951	1.3258940221	5.3437134181	38.00000000	38

table clearly manifests that at $r_c \to 0$, a 3DCHO has comparable $I_{\mathbf{r}}, I_{\mathbf{p}}$ values with that of PISB, thus confirming our presumption that, at $r_c \to 0$, 3DCHO behaves like a PISB. One also marks that, with reduction in r_c, $I_{\mathbf{p}}$ (and also potential energy) approaches zero. On the other hand, as r_c grows, the separation between $I_{\mathbf{r}}$ (also $I_{\mathbf{p}}$) values of 3DCHO and PISB tends to increase significantly. Moreover, as expected, at $r_c \to \infty$, 3DCHO reduces to 3DQHO. In previous section it was pointed out that, a 1DCHO may be treated as a two-mode system; at smaller (approaching zero) and larger (tending infinity) confinement lengths it behaving like a particle in a box and an 1DQHO respectively [180, 175]. Here, also we observe resembling behavior. At $r_c \to 0$ and ∞, 3DCHO leads to PISB and a 3D isotropic QHO respectively. We also note that, larger the value of ω higher will be $\langle v(r) \rangle$; as a matter of fact 3DCHO is more prone to 3DQHO in such a case. Conversely, lesser the ω value, $\langle v(r) \rangle$ is smaller and 3DCHO, in that occasion, is inclined towards a PISB. Therefore at a fixed r_c, by controlling ω values one can inquire the properties of all three systems starting from PISB to 3DQHO through 3DCHO.

So far, we have explored the limiting trend of 3DCHO. Now we look into its behavior at intermediate r_c region. For that, at first, the dependence of $I_{\mathbf{r}}, I_{\mathbf{p}}$ on quantum number n_r is recorded in Table 7. It tabulates these quantities for lowest five n_r (0-4) at six representative r_c values. This clearly implies that, at fixed m, l and r_c, both $I_{\mathbf{r}}, I_{\mathbf{p}}$ in 3DCHO get incremented as n_r attains higher values. Henceforth, the role of n_r on these measures is not discussed any further.

Table 8. $I_{\mathbf{r}}, I_{\mathbf{p}}$ of $1p, 1d, 1f$ states of 3DCHO at six r_c, with varying m. Last column correspond to respective 3DQHO values, computed from Eq. (77) [130]

$\|m\|$	$r_c = 0.1$	$r_c = 0.5$	$r_c = 1$	$r_c = 2$	$r_c = 7$	$r_c = \infty$
			$I_{\mathbf{r}}(1p)$			
0	8076.29142	323.0517082	80.76619765	20.39764116	10.00000000	10
1	5736.542528	229.4242339	57.21792995	13.91660078	6.00000000	6
			$I_{\mathbf{p}}(1p)$			
0	0.014984249	0.374503742	1.491857857	5.577935621	10.00000000	10
1	0.01003	0.2507	0.9980	3.975	6.0000	6
			$I_{\mathbf{r}}(1d)$			
0	13286.984765	531.4794285	132.8722779	33.37450170	14.00000000	14
1	10419.849672	416.7666252	104.0908347	25.74558754	10.00000000	10
2	7552.714578	302.0538220	75.3093915	18.11667339	6.00000000	6
			$I_{\mathbf{p}}(1d)$			
0	0.017547979	0.43862372	1.74993894	6.70582522	14.0000000	14
1	0.0133333	0.33326	1.32905	5.0595	10.00000	10
2	0.0091187	0.2279	0.9081	3.4132	6.00000	6
			$I_{\mathbf{r}}(1f)$			
0	19532.47745	781.2991270	195.3266196	48.95155358	18.000000000	18
1	16156.649498	646.2448005	161.48315319	40.15669405	14.00000000	14
2	12780.8215	511.1904739	127.63968669	31.36183453	10.00000000	10
3	9404.993581	376.1361474	93.79622020	22.566975007	6.000000000	6
			$I_{\mathbf{p}}(1f)$			
0	0.0194769435	0.486866098	1.944005783	7.551236509	18.0000000	18
1	0.0157907758	0.39471708	1.57572258	6.0995039	14.00000	14
2	0.012104608	0.30256807	1.20743938	4.6477714	10.00000	10
3	0.008418440	0.2104190	0.8391561	3.196038	6.00000	6

Now, in order to get a clear picture of the effect of magnetic quantum number, Table 8 gives $I_{\mathbf{r}}, I_{\mathbf{p}}$ of 3DCHO for lowest three nodeless states having $l \neq 0$, i.e., $1p$-$1f$ respectively, for all allowed m at 6 carefully selected r_c

Table 9. I_t for $1p$-$1f$ ($n_r = 0$) orbitals at six different r_c [130]. See text for details

$\|m\|$	$r_c = 0.1$	$r_c = 0.5$	$r_c = 1$	$r_c = 2$	$r_c = 7$	$r_c = \infty$
			$I_t(1p)^a$			
$0^\dagger$	121.01515	120.98286	120.491051	113.776614	100.00000000	100
1	57.53752	57.5166	57.1034	55.318	36.0000	36
			$I_t(1d)^b$			
$0^\dagger$	233.15973	233.117506	232.518248	223.80357	196.000000000	196
1	138.93098	138.89164	138.3419	130.2598	100.00000	100
2	68.870938	68.8380	68.3884	61.8358	36.00000	36
			$I_t(1f)^c$			
$0^\dagger$	380.432960	380.388051	379.716077	369.644758	324.0000000	324
1	255.126029	255.083860	254.452650	244.935911	196.00000	196
2	154.706834	154.669915	154.117184	145.762637	100.00000	100
3	79.1753741	79.1461919	78.7096703	72.1249096	36.00000	36

† These also correspond to upper bounds, given in Eq. (66).

^{a} Lower bounds Eq. (11), for $1p$ at 6 r_c are: 10.70940, 10.71226, 10.75598, 11.39074,12.96, 12.96.

^{b} Lower bounds Eq. (11), for $1d$ at 6 r_c are: 5.55842, 5.559428, 5.573756, 5.79079, 6.61224, 6.61224.

^{c} Lower bounds Eq. (11), for $1f$ at 6 r_c are: 3.40664, 3.40704, 3.41307, 3.506068, 4, 4.

$(0.1, 0.5, 1, 2, 7, \infty)$. Likewise, Figure 10 portrays these for $1g$ state in panels (a) and (b), for all possible $|m|$. It is to be noted that, the quantities in last column at $r_c = \infty$ are given from Eq. (77) considering $\omega = 1$. For this special ω, Eq. (77) dictates that, $I_{\mathbf{r}} = I_{\mathbf{p}}$ for 3DQHO. One notices that, behavior of $I_{\mathbf{r}}$ and $I_{\mathbf{p}}$ in 3DCHO is always in agreement with 3DQHO; an explicit analysis suggests identical patterns. In general, $I_{\mathbf{r}}$ decays monotonically while $I_{\mathbf{p}}$ progresses with rise in r_c; this is found to hold true for all $|m|$. This is to be expected, as a growth in r_c promotes delocalization in r space and localization in p space. At a fixed r_c and fixed quantum numbers n_r and m, $I_{\mathbf{r}}$ tends to rise with l; this is consistent to what is observed from Eq. (77) in 3DQHO. Because, in both 3DCHO and 3DQHO, kinetic energy grows with n_r, l. Similarly, at fixed n_r, l, in both 3DCHO and 3DQHO, $I_{\mathbf{r}}$ falls down as one descends down the table (increment in $|m|$). One striking fact is that, for all these three states considered, both 3DCHO and 3DQHO portray comparable pattern with respect to m. Next, I_t for $1p$-$1f$ states of 3DCHO are offered in Table 9. It seems to show an inclination towards $I_{\mathbf{r}}$ in the behavioral pattern. In all instances, the lower and upper bounds stated by Eq. (67) is satisfied. For an arbitrary state distinguished by quantum numbers n_r, l, change in $I_{\mathbf{r}}, I_{\mathbf{p}}$ with r_c in a 3DCHO preserves same qualitative orderings for various m (general nature of the plots remain unchanged) as depicted in these figures. This has been accomplished in a number of occasions, which are not reported here to save space. As usual, in all cases, they all eventually reach their respective 3DCHO-limit at some sufficiently large r_c, which alters from state to state.

Now to understand the dependence of $I_{\mathbf{r}}, I_{\mathbf{p}}$ on l, we present Table 10, where these are given for $l = 1 - 9$ states having $|m| = 1$ and n_r corresponding to 1, at same six chosen r_c values of previous table. The last column again has same significance as Table 8. In accordance with Eq. (77), here also for $n_r l$ ($n_r = 2$) states, $I_{\mathbf{r}} = I_{\mathbf{p}}$ in 3DQHO. Dependence of $I_{\mathbf{r}}, I_{\mathbf{p}}$ of 3DCHO on l compliments our observation in 3DQHO. At a fixed n_r, m and r_c, they both enhance with rise of l in 3DCHO and 3DQHO. This may occur probably because that, as l advances, the density gets increasingly concentrated. Therefore, at a certain n_r, m, a state with higher l undergoes greater fluctuation. Thus all our foregoing discussion leads to a general fact that,in a 3DCHO, the qualitative variations of $I_{\mathbf{r}}$ with all three quantum numbers remain quite analogous to that of $I_{\mathbf{p}}$; also the patterns in 3DCHO and 3DQHO are similar. It may be appropriate to mention a parallel work [129] along this direction for free and confined H atom inside an impenetrable spherical environment. One finds various significant deviations in the variation pattern between two systems there. Here we mention two of the most interesting facts, which are in complete contrast with a 3DCHO, e.g., $I_{\mathbf{r}}$ remains unchanged with respect to changes in l, whereas under confinement, it reduces with l (at fixed n, m). Besides, for a given state having fixed n, m, $I_{\mathbf{p}}$ enhances when the atom is compressed, whereas declines in a free H atom.

At this juncture, a few words may be devoted to the influence of ω on $I_{\mathbf{r}}, I_{\mathbf{p}}$. In order to establish this, Figure 11 depicts plots of $I_{\mathbf{r}}$ and $I_{\mathbf{p}}$ against r_c, for 5 selected ω^2 $(1, 2, 4, 8, 32)$, in bottom ((a), (b)) and top ((c), (d)) panels. These are provided for $1p$ state; left and right panels characterize $|m| = 0$ and 1 respectively. Evidently at a fixed r_c, $I_{\mathbf{r}}$ grows and $I_{\mathbf{p}}$ decays with increment in ω. At a certain ω, dependence of these measures on n_r, l, m is similar to

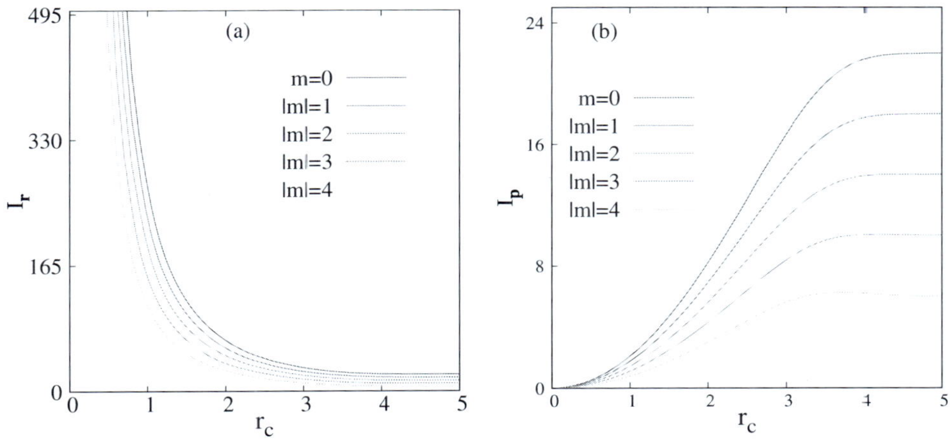

Figure 10. Variation of $I_{\mathbf{r}}$ and $I_{\mathbf{p}}$ in 3DCHO, with r_c, for all allowed $|m|$ values of $1g$ state, in panels (a) and (b) respectively [130]. See text for details.

that in 3DQHO. As ω goes up, there is more localization, hence compactness in single-particle density increases with oscillation strength.

An in-depth analysis of I reveals that, a gradual increase in r_c should lead to a delocalization of the system in such a way that, at $r_c \rightarrow \infty$ it should approach towards an 3DQHO. On the contrary, when $r_c \rightarrow 0$ the impact of confinement is maximum. It suffices for us to vary r_c from 0.1 to 8. This parametric increase in r_c elicits the system from an extremely confined environment to a free 3DQHO scenario.

Next, Table 11 shows evaluated $R_{\mathbf{r}}^{\alpha}$, $R_{\mathbf{p}}^{\beta}$, $R_t^{(\alpha,\beta)}$ for first two ($n_r = 1, 2$) s, p and d states of 3DCHO at a selected set of r_c values. $R_{\mathbf{r}}^{\alpha}$, starting from a $(-)$ve value at very low r_c, continuously increases, before finally reaching to the respective 3DQHO limit. This pattern in $R_{\mathbf{r}}^{\alpha}$ distinctly expresses delocalization of the system with growth in r_c. In contrast, $R_{\mathbf{p}}^{\beta}$, for all states generally tend to diminish monotonically with r_c, again converging to 3DQHO in the end. The analysis of $R_t^{(\alpha,\beta)}$ shows that, for $l = 0$ states it lowers with r_c to reach the unconfined result. However, for $1p$ state, it advances with r_c to attain the 3DQHO limit. In case of $2p$, $1d$ and $2d$ states it initially increases with r_c, then passes through a maximum, and finally convenes to limiting 3DQHO result. At small r_c (up to ≈ 2), $n_r = 1$ states have higher $R_{\mathbf{r}}^{\alpha}$ compared to their $n_r = 2$ counterparts. But at moderate r_c (> 2) region the situation alters; the $n_r = 2$ states now possess larger $R_{\mathbf{r}}^{\alpha}$. Moreover this crossover region appears at larger r_c with rise

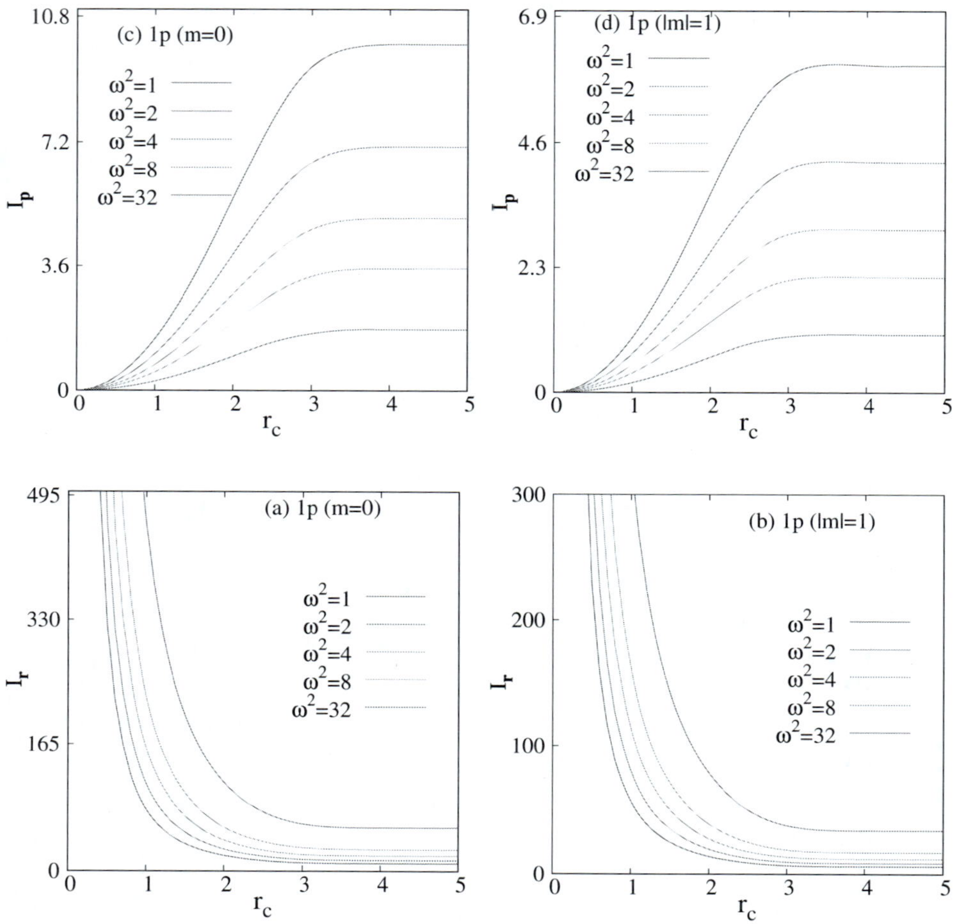

Figure 11. Plots of $I_{\mathbf{r}}$ and $I_{\mathbf{p}}$ in bottom ((a), (b)) and top ((c),(d)) panels in 3DCHO with r_c, of $1p$ state at five selected $\omega^2(1, 2, 4, 8, 32)$. Left, right columns correspond to $|m| = 0$, 1 respectively [130].

in l. This observation concludes that, at smaller r_c region the effect of confinement is more pronounced for higher n_r states (as $n_r = 1$ states possess higher $R_{\mathbf{r}}^{\alpha}$ values than that of respective $n_r = 2$ states). There is no such crossover in $R_{\mathbf{p}}^{\beta}$ and $R_t^{(\alpha,\beta)}$ in any of these reported states. The above observations are graphically demonstrated in Figure 12, where in segments (a)-(c), $R_{\mathbf{r}}^{\alpha}$, $R_{\mathbf{p}}^{\beta}$, $R_t^{(\alpha,\beta)}$ of first five circular states are presented as function of r_c. Panel (a) leads to the fact that, for all of them, $R_{\mathbf{r}}^{\alpha}$'s quite steadily progress with r_c and finally converge

Table 10. $I_{\mathbf{r}}$, $I_{\mathbf{p}}$ **for** $2p$-$2m$ ($|m| = 1$) **states in 3DCHO, at six selected** r_c **[130]**

l	$r_c = 0.1$	$r_c = 0.5$	$r_c = 1$	$r_c = 2$	$r_c = 7$	$r_c = \infty$
			$I_{\mathbf{r}}$			
1	19544.75146	781.7758499	195.3902144	48.61008407	14.000000000	14
2	28201.56829	1128.0488516	281.9599554	70.272141773	18.000000000	18
3	37976.00550	1519.027385	379.70865634	94.73419843	22.0000000000	22
4	48838.86407	1953.5428785	488.3419262	121.91517166	26.0000000000	26
5	60766.49285	2430.649110	607.62258270	151.75456390	30.0000000001	30
6	73739.4732	2949.5692932	737.3562778	184.204055587	34.0000000743	34
7	87741.58879	3509.6547640	877.38084537	219.22364350	38.00000037	38
8	102759.07974	4110.3551463	1027.5587357	256.77944910	42.0000017797	42
9	118780.106607	4751.196872	1187.77161130	296.84230573	46.0000074341	46
			$I_{\mathbf{p}}$			
1	0.01221	0.3054	1.222	4.917	14.000	14
2	0.0133333	0.3333337	1.33356	5.34750	18.00000	18
3	0.014439157	0.36097833	1.4438781	5.7743588	21.999997	22
4	0.0154743767	0.38685628	1.54723822	6.1782594	26.0000000	26
5	0.016426161	0.4106495	1.6423270	6.55303527	30.000000	30
6	0.017296672	0.4324115849	1.7293332625	6.898114452	34.00000	34
7	0.0180927445	0.4523131126	1.808922468	7.21517631	38.00000	38
8	0.0188222240	0.470550071	1.88186836	7.50667165	41.9999991	42
9	0.019492648	0.487310795	1.94891804	7.775182	45.99998	46

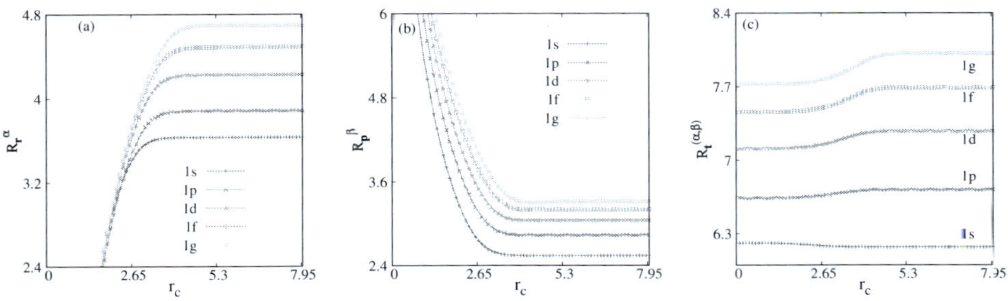

Figure 12. Plots of $R_{\mathbf{r}}^{\alpha}$, $R_{\mathbf{p}}^{\beta}$, $R_t^{(\alpha,\beta)}$ against r_c for first five circular states of 3DCHO in panels (a), (b), (c) respectively [181]. See text for details.

to 3DQHO. Similarly, from panel (b) it is clear that, $R_{\mathbf{p}}^{\beta}$ shows reverse pattern with r_c, before reaching 3DQHO-limit. From the last panel, it appears that, for $l = 0$ state, $R_t^{(\alpha,\beta)}$ abates with r_c, whereas an opposite trend occurs otherwise. Also one finds that major changes in $R_t^{(\alpha,\beta)}$ occurs in the range of r_c varying from 2.65-5.30, with the higher state showing most predominant effect (see Figures 15 and 17 later for S_t and E_t, for similar effects). Some of these may not be evident from the figure as such, but becomes clear upon magnification of the plot, in consultation with the data given in Table 11.

To gain further insight, Figure 13 delineates $R_{\mathbf{r}}^{\alpha}$, $R_{\mathbf{p}}^{\beta}$, $R_t^{(\alpha,\beta)}$, in left (a), middle (b), right (c) panels, for lowest five circular states with changes in n_r (maximum being 9). Four different r_c's are chosen, namely, $0.1, 3, 5, \infty$ in segments (A)-(D) from bottom to top. At the lowest r_c, for all l, $R_{\mathbf{r}}^{\alpha}$'s gradually falls off with advancement in n_r, providing highest values for $l = 0$. But for *non-zero* l, it increases with l for smaller n_r. However, at $n_r > 6$, there seems to be a crossover between $l = 1$ and $l = 2$ states. Hence it can be concluded that, effect of confinement is maximum, minimum for $l = 1$, $l = 0$ states respectively. At a fixed l and r_c (lower region), higher n_r-states encounter the confinement to a greater extent. On the other hand, both $R_{\mathbf{p}}^{\beta}$ (b) and $R_t^{(\alpha,\beta)}$ (c) show inverse trend. For smaller n_r, both these quantities follow the sequence $1g > 1f > 1d > 1p > 1s$, which however, gets reversed ($1s > 1p > 1d > 1f > 1g$) at higher n_r. These results infers that, at lower r_c region quantum effect gets amplified as information content decreases, whereas total information increases with n_r. The first column (a) interestingly shows the appearance of maximum in $R_{\mathbf{r}}^{\alpha}$ at some moderate r_c, whose positions get right shifted as r_c enhances. Apparently there exists an interplay between two opposing effects: (i) radial confinement (localization) (ii) accumulation of nodes with n_r (delocalization). As r_c progresses, delocalization prevails for lower n_r. Hence with gradual relaxation in confinement, states with larger n_r get delocalized. At $r_c \rightarrow \infty$, the second effect predominates; so the system behaves like 3DQHO. In second and third columns, one sees that, both $R_{\mathbf{p}}^{\beta}$, $R_t^{(\alpha,\beta)}$ enhance with n_r. At $r_c \rightarrow \infty$, these two quantities approach the respective 3DQHO-limit, as predicted.

Next we move to S in Table 12, where $S_{\mathbf{r}}$, $S_{\mathbf{p}}$ and S_t are demonstrated for $1s, 2s, 1p, 2p, 1d, 2d$ states of 3DCHO at same set of r_c of the previous table. Akin to $R_{\mathbf{r}}^{\alpha}$, $S_{\mathbf{r}}$ also yield $(-)$ve values for all six states at very low r_c, and then continuously increase until reaching the limiting 3DQHO value. This trend

Table 11. $R^{\alpha}_{\mathbf{r}}$, $R^{\beta}_{\mathbf{p}}$ and $R^{\alpha,\beta}_t$ for lowest three l (having $n_r = 1, 2$) states in 3DCHO at eight selected values of r_c. $R^{\alpha,\beta}_t$ for all these states obey the lower bound given in Eq. (57) [181]

r_c	$R^{\alpha}_{\mathbf{r}}$	$R^{\beta}_{\mathbf{p}}$	$R^{\alpha,\beta}_t$	r_c	$R^{\alpha}_{\mathbf{r}}$	$R^{\beta}_{\mathbf{p}}$	$R^{\alpha,\beta}_t$
	$1s$				$2s$		
0.1	−6.0366917844	12.24717[illegible]	6.210480	0.1	−6.0653334752	14.251561	8.186227
0.2	−3.9572613535	10.167740	6.210479	0.2	−3.9858896952	12.172113	8.185223
0.5	−1.2088403559	7.4192722	6.210431	0.5	−1.2369267194	9.422993	8.186067
1.0	0.8636306014	5.3460818	6.209712	1.0	0.8438971150	7.339578	8.183475
5.0	3.6326806767	2.5410652	6.173745	5.0	4.5764993107	2.616203	7.192702
8.0	3.6326909163	2.5410540	6.173744	8.0	4.5767695172	2.614825	7.191594
∞	3.6326909163	2.5410540	6.173744	∞	4.5767695172	2.614825	7.191594
	$1p$				$2p$		
0.1	−6.1671363542	12.806502	6.639365	0.1	−6.29011971	14.178463	7.888343
0.2	−4.0876997330	10.727065	6.639366	0.2	−4.21067738	12.099025	7.888348
0.5	−1.3390273792	7.978413	6.639386	0.5	−1.46177311	9.350298	7.888525
1.0	0.7373200936	5.902381	6.639701	1.0	0.61816051	7.273091	7.891251
5.0	3.8830108378	2.834951	6.717917	5.0	4.57588095	2.982808	7.558689
8.0	3.8830566606	2.834947	6.718004	8.0	4.57683522	2.982008	7.558343
∞	3.8830566606	2.834947	6.718004	∞	4.57683522	2.982008	7.558343
	$1d$				$2d$		
0.1	−6.1181191683	13.228954	7.110835	0.1	−6.2478841627	14.295117	8.047233
0.2	−4.0386800101	11.149516	7.110836	0.2	−4.1684423307	12.215678	8.047236
0.5	−1.2899046230	8.400780	7.110875	0.5	−1.4195583633	9.466934	8.047376
1.0	0.7880364400	6.323450	7.111487	1.0	0.6600654534	7.389472	8.049538
5.0	4.2284239258	3.048859	7.277282	5.0	4.8017869530	3.233068	8.034855
8.0	4.2285908429	3.047026	7.275616	8.0	4.8046032500	3.230664	8.035267
∞	4.2285908429	3.047026	7.275616	∞	4.8046032500	3.230664	8.035267

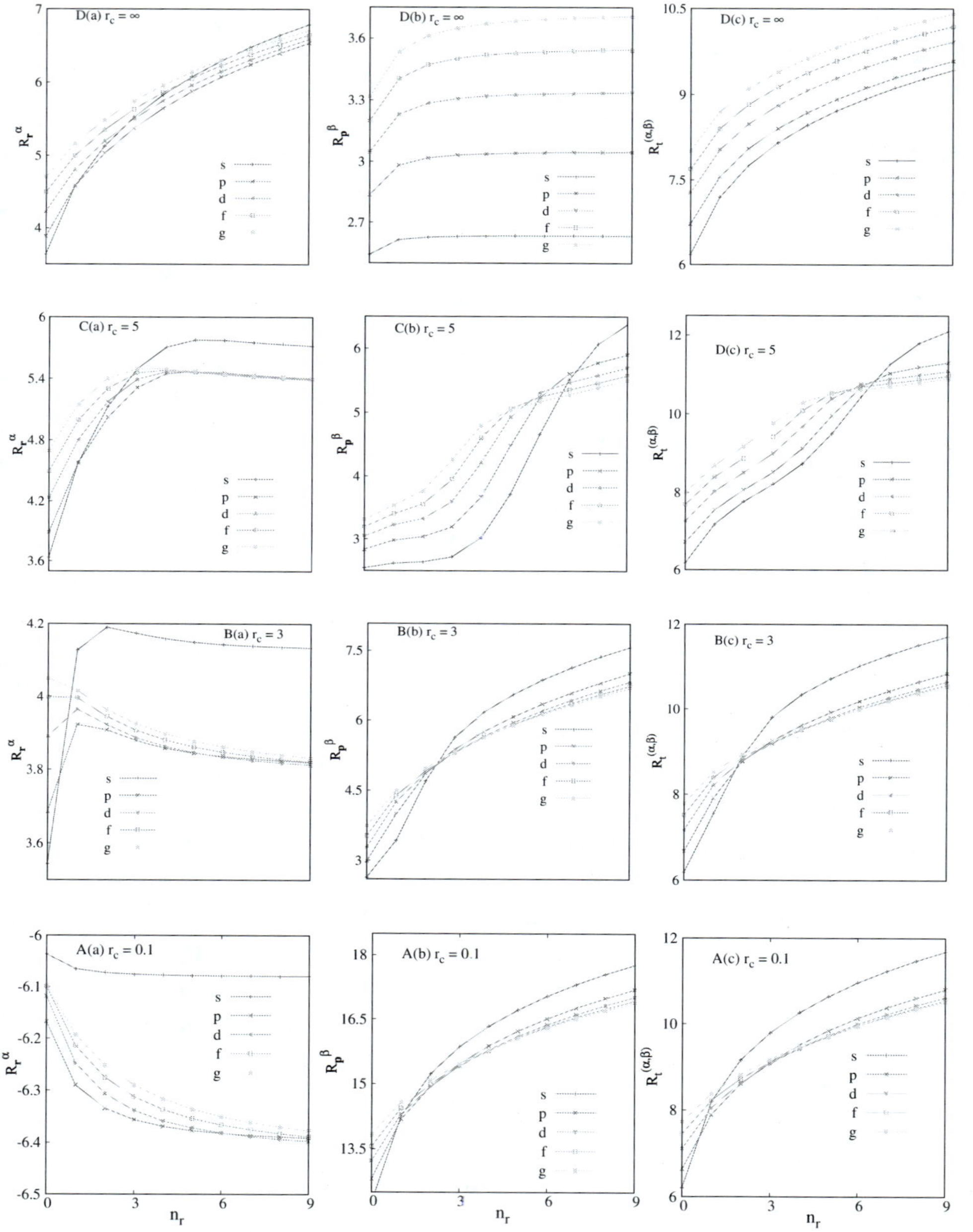

Figure 13. Plot of $R_{\mathbf{r}}^{\alpha}$ (a), $R_{\mathbf{p}}^{\beta}$ (b) and $R_t^{(\alpha,\beta)}$ (c) versus n_r (at $\omega = 1$) for s, p, d, f, g states at four particular r_c's of 3DCHO, namely, $0.1, 3, 5, \infty$ in panels (A)-(D). $R_t^{(\alpha,\beta)}$'s for all these states obey the lower bound given in Eq. (57) [181]. For more details, consult text.

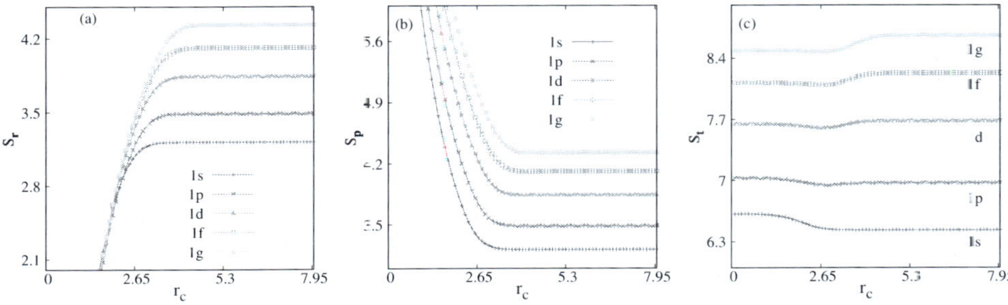

Figure 14. Plots of $S_{\mathbf{r}}$, $S_{\mathbf{p}}$, S_t against r_c for first five circular states of 3DCHO in panels (a), (b), (c) respectively [181]. See text for details.

of $S_{\mathbf{r}}$ also interprets the delocalization of 3DCHO with progress in r_c. Like $R^{\beta}_{\mathbf{p}}$, $S_{\mathbf{p}}$ also offers a reverse nature (continuous decrease with r_c) of $S_{\mathbf{r}}$ (and $R^{\alpha}_{\mathbf{r}}$); from an initial $(+)$ve value, consistently falls off to reach 3DQHO. S_t's for both $1s$, $1p$ states decrease to attain 3DQHO result. Interestingly, for $1d$ state it first declines, attains a minimum value, then rises to a maximum and ultimately reaches to 3DQHO values. But for $2s$, $2p$, $2d$ states it first increases, then reaches a maximum and finally merges to borderline 3DQHO values.

Figure 14 delineates behavioral trends of $S_{\mathbf{r}}$, $S_{\mathbf{p}}$, S_t with r_c, in segments (a)-(c), for same five states of Figure 12 (done for R). One notices that, both Figures 12 and 14 offer qualitatively analogous changes with respect to r_c variation; thus signalling comparable trends in R and S. For all these states $S_{\mathbf{r}}$'s tend to progress with r_c and then reach the corresponding r-space 3DQHO, while $S_{\mathbf{p}}$'s reduce before attaining that. The last panel portrays that, for $1s$, $1p$ states S_t decreases with r_c, while for the remaining three states, S_t passes through a minimum. It is worth noting that, in the $1p$ state, R_t progresses and S_t regresses after $r_c = 2.65$. In both occasions, however, as usual they all gain the limiting 3DQHO behavior at a sufficiently large r_c.

In Figure 15, $S_{\mathbf{r}}$ (a), $S_{\mathbf{p}}$ (b), S_t (c) of $l = 0-4$ states are plotted with respect to n_r at same four values of r_c in Figure 13, in panels marked with (A)-(D) from bottom to top. Again, these graphs produce resembling shape and propensity as in Figure 13. Thus in harmony with $R^{\alpha}_{\mathbf{r}}$ (for all five l at the lowest $r_c = 0.1$), $S_{\mathbf{r}}$ gets lowered in A(a), while $S_{\mathbf{p}}$s and S_ts accelerate with n_r in A(b) and A(c), respectively. This supports our previous observation (of R in Figure 13) that, at very low r_c, the effect of confinement is more prominent in high-lying states, signifying an intensification of quantum nature in such a scenario. As usual,

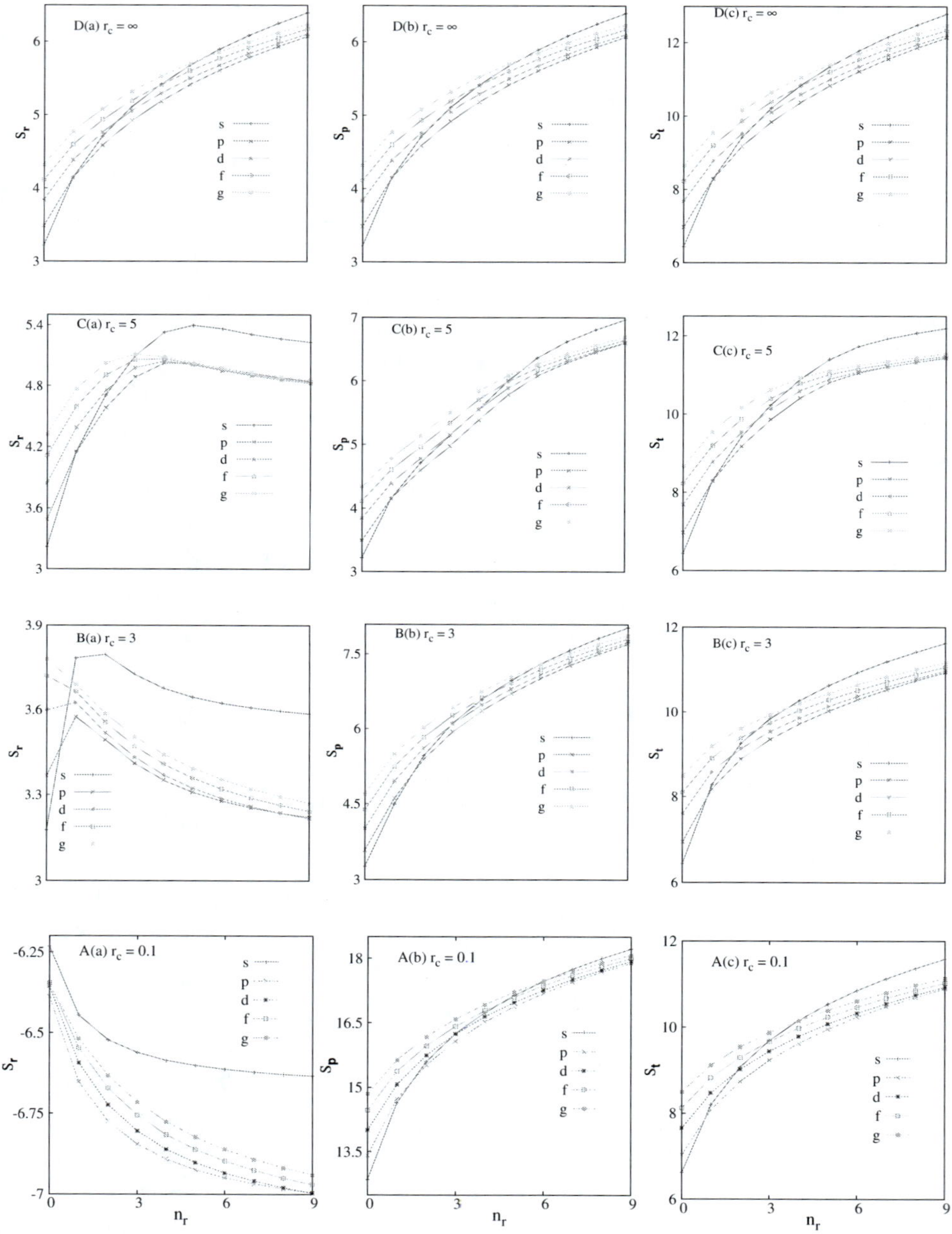

Figure 15. Plot of $S_{\mathbf{r}}$ (a), $S_{\mathbf{p}}$ (b) and S_t (c) versus n_r (at $\omega = 1$) for s, p, d, f, g states at four particular r_c's of 3DCHO, namely, $0.1, 3, 5, \infty$ in panels (A)-(D). S_t's for all these states obey the lower bound given in Eq. (53) [181]. For more details, consult text.

Table 12. $S_{\mathbf{r}}$, $S_{\mathbf{p}}$ and S_t for lowest three l (having $n_r = 1, 2$) states in 3DCHO at eight selected values of r_c. S_t for all these states obey the lower bound given in Eq. (53) [181]

r_c	$S_{\mathbf{r}}$	$S_{\mathbf{p}}$	S_t	r_c	$S_{\mathbf{r}}$	$S_{\mathbf{p}}$	S_t
	$1s$				$2s$		
0.1	−6.232173222	12.8494	6.6172	0.1	−6.4460987687	14.6389	8.1923
0.2	−4.152747179	10.7700	6.6172	0.2	−4.3666534417	12.5595	8.1928
0.5	−1.404504328	8.0214	6.6168	0.5	−1.6176276192	9.8106	8.1929
1.0	0.6652222004	5.9458	6.6110	1.0	0.4641636149	7.731	8.195
5.0	3.2170947394	3.217094	6.434189	5.0	4.150729546	4.1510	8.3017
8.0	3.2170948239	3.217094	6.434189	8.0	4.150745547	4.15074	8.30148
∞	3.2170948239	3.217094	6.434189	∞	4.150745547	4.15074	8.30148
	$1p$				$2p$		
0.1	−6.38738206	13.418	7.030	0.1	−6.651966568	14.7283	8.0763
0.2	−4.30794705	11.338	7.030	0.2	−4.572523919	12.6489	8.0763
0.5	−1.55934019	8.5894	7.0300	0.5	−1.823606736	9.9000	8.0763
1.0	0.51599338	6.5114	7.0273	1.0	0.256528223	7.82132	8.07784
5.0	3.4874566574	3.4875	6.9750	5.0	4.1477548396	4.1483	8.2960
8.0	3.4874576660	3.487457	6.974915	8.0	4.14786196159	4.147861	8.295723
∞	3.4874576660	3.487457	6.974915	∞	4.14786196159	4.147861	8.295723
	$1d$				$2d$		
0.1	−6.3553068427	14.0035	7.6481	0.1	−6.5939939435	15.0676	8.4736
0.2	−4.2758683878	11.9241	7.6481	0.2	−4.5145520348	12.9881	8.4736
0.5	−1.52712156	9.1753	7.6481	0.5	−1.7656649279	10.2393	8.4736
1.0	0.5503764295	7.0964	7.6467	1.0	0.3140078818	8.1605	8.4745
5.0	3.8426303929	3.8427	7.6853	5.0	4.3885945973	4.3909	8.7794
8.0	3.8426381378	3.842638	7.685276	8.0	4.389113529281	4.389113	8.773227
∞	3.8426381378	3.842638	7.685276	∞	4.389113529281	4.389113	8.773227

like $R_{\mathbf{r}}^{\alpha}$ here also, the first column (a) shows the appearance of maximum in $S_{\mathbf{r}}$ plot with gradual progress of r_c from the bottom to top. Their positions switch to right as r_c enhances. This observation indicates that, at $r_c \rightarrow \infty$ the system behaves like 3DQHO.

At this point we proceed for the last measure (E) of this study in Table 13. A cross-section of $E_{\mathbf{r}}$, $E_{\mathbf{p}}$, E_t for $1s, 2s, 1p, 2p, 1d, 2d$ states of 3DCHO (same

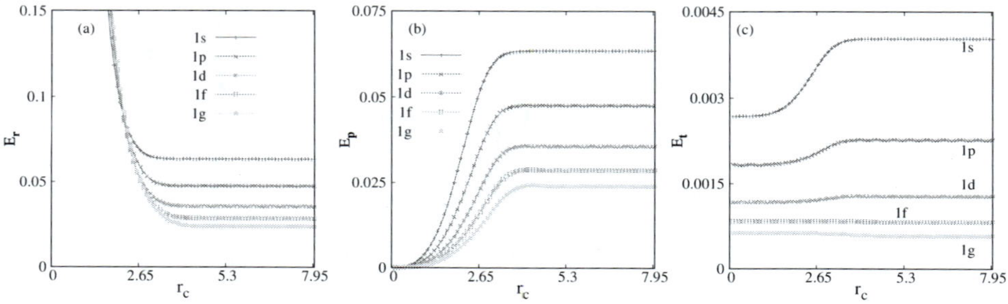

Figure 16. Plots of $E_{\mathbf{r}}$, $E_{\mathbf{p}}$, E_t against r_c for first five circular states of 3DCHO in panels (a), (b), (c) respectively [181]. See text for details.

set of r_c as used in previous table) is offered. One notices that, $E_{\mathbf{r}}$ reduces (as opposed to $R_{\mathbf{r}}^{\alpha}$, $S_{\mathbf{r}}$) while $E_{\mathbf{p}}$ progresses (in contrast to $R_{\mathbf{p}}^{\beta}$, $S_{\mathbf{p}}$) with rise in r_c. However, the nature of variation of E_t with r_c varies from state to state. For $1s, 1p, 1d$ states, it enhances with r_c, while $2s, 2p, 2d$ states, it goes through a minimum. These changes in $E_{\mathbf{r}}$, $E_{\mathbf{p}}$ and E_t with r_c are demonstrated pictorially in Figure 16, in left (a), middle (b) and right (c) panels for first five node-less states. It is found that, E_t for $1s$, $1p$, $1d$ states advances with r_c, while for $1f$, $1g$ states it decreases. Interestingly, at large r_c, both $E_{\mathbf{r}}$, $E_{\mathbf{p}}$ decline with growth in l. Finally, Figure 17, depicts $E_{\mathbf{r}}$, $E_{\mathbf{p}}$, E_t (in columns (a),(b),(c)) for $l = 0 - 4$ states as functions of n_r at four different r_c (in segments A-D). At the lowest r_c considered, $E_{\mathbf{r}}$ advances with n_r. However, the first column (a) suggests that, a minimum appears in $E_{\mathbf{r}}$ graphs as r_c is increased. Also the positions of these minima shift towards right with rise in r_c. On the contrary, for all concerned r_c, both $E_{\mathbf{p}}$, E_t decay with n_r.

All the discussed measures of Eq. (72) have been investigated with change of r_c, choosing two different and widely used values of b $(1, \frac{2}{3})$. Note that, for $b = 1$, $C_{ES}^{(2)}$ transforms to C_{LMC}; similarly $C_{IS}^{(1)}$ coincides with C_{IS} at $b = \frac{2}{3}$. In order to simplify the discussion, a few words may be devoted to the notation used. A uniform symbol $C^{b}_{order_s, disorder_s}$ is used; where the two subscripts refer to two order (E, I) and disorder (S, R) parameters. Another subscript s is used to specify the space; *viz.*, r, p or t (total). Two scaling parameters $b = \frac{2}{3}, 1$ are identified with superscripts (1), (2). These measures are examined systematically for $1s$, $1p$, $1d$, $2s$, $1f$, $2p$, $1g$ and $2d$ states in conjugate spaces. In this part we will discuss the outcome extracted from a recent complexity analysis undertaken by us. The detailed results with relevant figures and tables

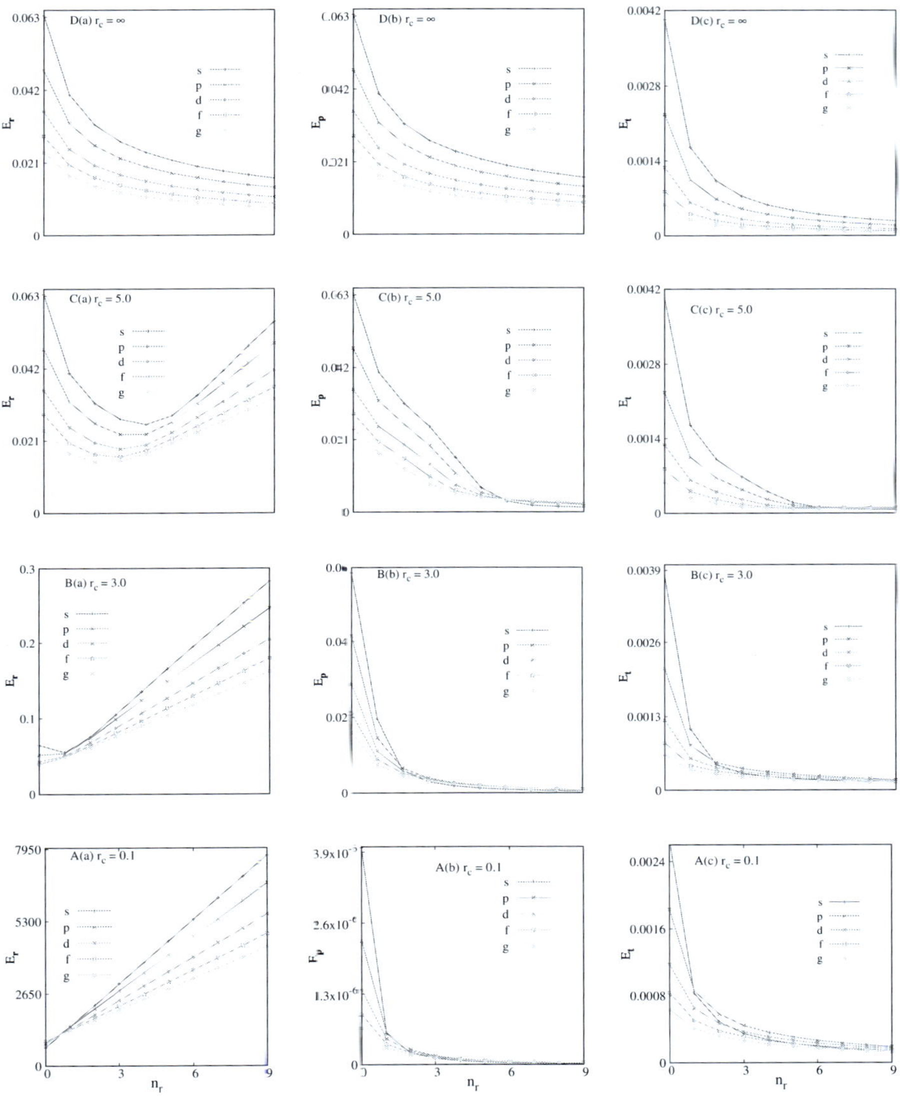

Figure 17. Plot of $E_{\mathbf{r}}$ (a), $E_{\mathbf{p}}$ (b) and E_t (c) versus n_r (at $\omega = 1$) for s, p, d, f, g states at four particular r_c's of 3DCHO, namely, $0.1, 3, 5, \infty$ in panels (A)-(D) [181]. For more details, consult text.

will be published elsewhere separately.

Now we lay our focus on $C^{(1)}_{E_{\mathbf{r}}S_{\mathbf{r}}}, C^{(1)}_{E_{\mathbf{p}}S_{\mathbf{p}}}$ and $C^{(2)}_{E_{\mathbf{r}}S_{\mathbf{r}}}, C^{(2)}_{E_{\mathbf{p}}S_{\mathbf{p}}}$ respectively. $C^{(1)}_{E_{\mathbf{r}}S_{\mathbf{r}}}$ decreases and $C^{(1)}_{E_{\mathbf{p}}S_{\mathbf{p}}}$ increases with rise of r_c, before reaching the borderline 3DQHO result. As mentioned earlier, the 3DCHO model acts as an intermediate between the PISB and 3DQHO [130]. That means, an enhancement in r_c should lead the system towards delocalization (equilibrium), which is supported by the declining nature of $C^{(1)}_{E_{\mathbf{r}}S_{\mathbf{r}}}$ with progress in r_c. At a fixed l, $C^{(1)}_{E_{\mathbf{r}}S_{\mathbf{r}}}$ progresses with n_r. It may be noted that in a 3DCHO, number of nodes varies with n_r only (in contrast to, say the H atom, where it depends on both $n = n_r$ and l). Hence in this case, the increment in number of nodes takes the system towards the side of ordering. But no such pattern is observed for $C^{(1)}_{E_{\mathbf{p}}S_{\mathbf{p}}}$. The variation of $C^{(2)}_{E_{\mathbf{r}}S_{\mathbf{r}}}$, $C^{(2)}_{E_{\mathbf{p}}S_{\mathbf{p}}}$ with change of r_c reveals that, former advances with growth of r_c indicating that this is more inclined towards order. On the other hand, at first $C^{(2)}_{E_{\mathbf{p}}S_{\mathbf{p}}}$ decreases with r_c, attains a minimum and finally merges to 3DQHO. The ordering of $C^{(2)}_{E_{\mathbf{r}}S_{\mathbf{r}}}$ regarding n_r quantum number is akin to $C^{(1)}_{E_{\mathbf{r}}S_{\mathbf{r}}}$. This observation complements our previous conjecture that, an increase in number of nodes takes the system towards disequilibrium. It is noticed that, after a certain r_c (≈ 3), both $C^{(2)}_{E_{\mathbf{r}}S_{\mathbf{r}}}, C^{(2)}_{E_{\mathbf{p}}S_{\mathbf{p}}}$ show analogous nature (increase to reach their respective limiting value). Here, exploring these two sets of complexity measure, namely, $C^{(1)}_{ES}$ and $C^{(2)}_{ES}$, it is evident that, $C^{(1)}_{E_{\mathbf{r}}S_{\mathbf{r}}}$ and $C^{(1)}_{E_{\mathbf{p}}S_{\mathbf{p}}}$ complement each other better (as former decreases, later increases with r_c) and more tends towards disorder (equilibrium), whereas the other set is inclined towards the order-direction. Thus, $C^{(1)}_{ES}$ with $b = \frac{2}{3}$ explains the system better. It is important to point out that, when 3DCHO approaches to 3DQHO, $C^{(1)}_{E_{\mathbf{r}}S_{\mathbf{r}}}$ becomes equal to $C^{(1)}_{E_{\mathbf{p}}S_{\mathbf{p}}}$.

In the same direction, now we interpret the pattern changes of $C^{(1)}_{E_{\mathbf{r}}R_{\mathbf{r}}}, C^{(1)}_{E_{\mathbf{p}}R_{\mathbf{p}}}$ with r_c. At first, $C^{(1)}_{E_{\mathbf{r}}R_{\mathbf{r}}}$ declines with growth of r_c, then passes through a minimum and finally converges to respective 3DQHO result. This minimum gets flatter with rise of both n_r, l. Here also $C^{(1)}_{E_{\mathbf{r}}R_{\mathbf{r}}}$ shows similar trend as observed in $C^{(1)}_{E_{\mathbf{r}}S_{\mathbf{r}}}$, *viz*., (i) like $C^{(1)}_{E_{\mathbf{r}}S_{\mathbf{r}}}$, $C^{(1)}_{E_{\mathbf{r}}R_{\mathbf{r}}}$ is more aligned towards disorder (ii) at a particular l, they progress with n_r. The first point suggests that, like $C^{(1)}_{E_{\mathbf{r}}S_{\mathbf{r}}}$, $C^{(1)}_{E_{\mathbf{r}}R_{\mathbf{r}}}$ also interprets that, with increase in r_c the system progresses towards 3DQHO from PISB via 3DCHO. The second point says that, an increase/decrease in number of nodes leads towards the direction of order/disorder respectively. Like $C^{(1)}_{E_\mathbf{p}S_\mathbf{p}}$, $C^{(1)}_{E_\mathbf{p}R_\mathbf{p}}$ increases with growth in r_c. Moreover, at a definite l, $C^{(1)}_{E_\mathbf{p}R_\mathbf{p}}$ falls off with n_r. At smaller r_c region (≈ 3), both $C^{(2)}_{E_\mathbf{r}R_\mathbf{r}}$, $C^{(2)}_{E_\mathbf{p}R_\mathbf{p}}$ vary slowly. At around same r_c, the former jumps and latter drops to attain the borderline 3DQHO limit. For both $C^{(2)}_{E_\mathbf{r}R_\mathbf{r}}$ and $C^{(2)}_{E_\mathbf{p}R_\mathbf{p}}$ absolute values of the slope of the curve enhance as n_r grows (fixed l) and decrease with growth of l (fixed n_r). The dependence of $C^{(2)}_{E_\mathbf{r}R_\mathbf{r}}$ on n_r is similar to $C^{(1)}_{E_\mathbf{r}R_\mathbf{r}}$. This again validates the dependency of this complexity measure on number of nodes of the system. Once again one may conclude that, out of $C^{(1)}_{ER}$ and $C^{(2)}_{ER}$, the former offers a more complete knowledge about the characteristics of 3DCHO. Apart from that, $C^{(1)}_{E_\mathbf{p}R_\mathbf{p}}$ shows more lucid trend than $C^{(1)}_{E_\mathbf{p}S_\mathbf{p}}$ with respect to dependence of quantum number n_r. Hence, in practice, $C^{(1)}_{ER}$ may possibly be considered a better measure of complexity than $C^{(1)}_{ES}$.

Now we discuss variations of $\{C^{(1)}_{I_\mathbf{r}S_\mathbf{r}}, C^{(1)}_{I_\mathbf{p}S_\mathbf{p}}\}$ and $\{C^{(2)}_{I_\mathbf{r}S_\mathbf{r}}, C^{(2)}_{I_\mathbf{p}S_\mathbf{p}}\}$ with changes in r_c for all the states mentioned above. Interestingly. in this occasion, the nature of alteration of $\{C^{(1)}_{I_\mathbf{r}S_\mathbf{r}}, C^{(1)}_{I_\mathbf{p}S_\mathbf{p}}\}$ with r_c varies from state to state. At a fixed n_r, both $C^{(1)}_{I_\mathbf{r}S_\mathbf{r}}$, $C^{(1)}_{I_\mathbf{p}S_\mathbf{p}}$ increase with l. Further, at a fixed l, they elevate with growth in n_r. This behavior of $C^{(1)}_{I_\mathbf{r}S_\mathbf{r}}$ with n_r, is again connected to the nodal structure in C_{ES} and C_{ER}. On the contrary, $C^{(2)}_{I_\mathbf{r}S_\mathbf{r}}$, $C^{(2)}_{I_\mathbf{p}S_\mathbf{p}}$ increase and decrease with growth in r_c and finally approach to respective 3DQHO. This obviously conveys that, $C^{(2)}_{I_\mathbf{r}S_\mathbf{r}}$ is a descriptor of disequilibrium. At a particular n_r both $C^{(2)}_{I_\mathbf{r}S_\mathbf{r}}$, $C^{(2)}_{I_\mathbf{p}S_\mathbf{p}}$ enhance with advancement in l. Also, at a certain l, they enhance with improvement of n_r. However, this pattern is in agreement with C_{ES} and C_{ER}. In case of 3DCHO, both $\{C^{(1)}_{I_\mathbf{r}S_\mathbf{r}}, C^{(1)}_{I_\mathbf{p}S_\mathbf{p}}\}$ and $\{C^{(2)}_{I_\mathbf{r}S_\mathbf{r}}, C^{(2)}_{I_\mathbf{p}S_\mathbf{p}}\}$ are unable to explain the delocalization of the system from PISB to 3DQHO through the increase in r_c.

Finally, we mention a few words regarding the behavioral pattern in our last complexity measure, *viz.*, $C^{(1)}_{I_\mathbf{r}R_\mathbf{r}}$, $C^{(1)}_{I_\mathbf{p}R_\mathbf{p}}$, $C^{(2)}_{I_\mathbf{r}R_\mathbf{r}}$, $C^{(2)}_{I_\mathbf{p}R_\mathbf{p}}$, with alterations in r_c. It is found that, $C^{(1)}_{I_\mathbf{r}R_\mathbf{r}}$ progresses slowly with r_c to reach respective 3DQHO values, implying that, $C^{(1)}_{I_\mathbf{r}R_\mathbf{r}}$ is an order parameter. At a suitable n_r, this quantity advances with l. In a parallel way, at constant l, $C^{(1)}_{I_\mathbf{r}R_\mathbf{r}}$ accumulates with n_r. This again means that with increase in number of nodes, the system moves towards order. Besides this, it can be concluded that, for node-less (like $1s, 1p, 1d, 1f, 1g$) states, $C^{(1)}_{I_\mathrm{p}R_\mathrm{p}}$ diminishes with progression in r_c. But for states having one radial node (such as $2s, 2p, 2d$), there appears a maximum in $C^{(1)}_{I_\mathrm{p}R_\mathrm{p}}$, which gets right shifted with increase in l. It can be said that, $C^{(2)}_{I_\mathrm{r}R_\mathrm{r}}$, $C^{(2)}_{I_\mathrm{p}R_\mathrm{p}}$, accelerate and decelerate respectively with growth in r_c. Hence, $C^{(2)}_{IR}$ is an indicator of order. Like $C^{(1)}_{IR}$, both $C^{(2)}_{I_\mathrm{r}R_\mathrm{r}}$, $C^{(2)}_{I_\mathrm{p}R_\mathrm{p}}$ enhance with emergence of n_r and l quantum numbers. A closer investigation conveys that, neither $C^{(1)}_{IR}$ nor $C^{(2)}_{IR}$ can explain the transformation of system from PISB to 3DQHO via 3DCHO. It is hoped that, this study would be useful for future references and would inspires further work along this direction.

4.3. Relative Information

4.3.1. 1D Quantum Harmonic Oscillator (1DQHO)

At first we would like to explore IR in model 1DQHO in x, p spaces. The underlying potential is characterized by the expression: $v(x) = \frac{1}{2}\omega^2 x^2$ (mass m is set to unity throughout), where ω signifies angular frequency. The normalized x-space wave function is expressed as ($H_n(x)$ refers to Hermite polynomial),

$$\psi_n(x) = \left(\frac{\omega}{\sqrt{2}\pi}\right)^{\frac{1}{4}} \frac{1}{\sqrt{2^n n!}}\, H_n\left(\frac{\sqrt{\omega}}{2^{\frac{1}{4}}}x\right) e^{-\frac{\omega}{2\sqrt{2}}x^2}. \tag{80}$$

Choosing $n = 0$ as reference, $\frac{\sqrt{\omega}}{2^{\frac{1}{4}}}x = y$, and using definition of IR in Eq. (73), one gets,

$$\mathrm{IR}_x = \frac{\omega}{\sqrt{2}}\,\frac{1}{2^{n-1}n!}\int_0^\infty H_n^2(y)\, e^{-y^2}\left[\frac{\psi_n'(y)}{\psi_n(y)} - \frac{\psi_0'(y)}{\psi_0(y)}\right]^2 \mathrm{d}y. \tag{81}$$

Here the suffix "x" denotes a position-space quantity. Now use of recurrence relation $H_n'(y) = 2nH_{n-1}(y)$, in conjunction with orthonormality condition of Hermite polynomial, $\int_0^\infty e^{-y^2} H_m(y)H_k(y)\mathrm{d}y = 2^{m-1}m!\sqrt{\pi}\delta_{mk}$ ($\delta_{mn} = 1$ when $m = k$, and 0 otherwise) produces,

$$\mathrm{IR}_x = 4n^2\frac{\omega}{\sqrt{2}}\,\frac{1}{2^{n-1}n!}\int_0^\infty [H_{n-1}(y)]^2\, e^{-y^2}\mathrm{d}y = 4\sqrt{2}\,\omega n. \tag{82}$$

Thus Eq. (82) suggests that, IR_x in nth state may be obtained from a knowledge of $(n-1)$th-state wave function. Clearly, it increases linearly with state index

n, with a positive slope of $4\sqrt{2}\omega$. This is in agreement with the fact that in this system, localization as well as fluctuation enhance with ω. So IR_x result simply complements this.

Now we move on to p space, where the normalized wave function is in the form of,

$$\psi_n(p) = \left(\frac{\sqrt{2}}{\omega\pi}\right)^{\frac{1}{4}} \frac{1}{\sqrt{2^n n!}} H_n\left(\frac{2^{\frac{1}{4}}}{\sqrt{\omega}}p\right) e^{-\frac{1}{\sqrt{2}\omega} p^2}. \tag{83}$$

Again, considering $n = 0$ state as reference, setting $\frac{2^{\frac{1}{4}}}{\sqrt{\omega}}p = g$, using the recurrence relation $H_n'(g) = 2nH_{n-1}(g)$ and invoking orthonormality condition of Hermite polynomial, one can derive the following expression for IR_p after some straightforward algebra (p subscript indicates p-space quantity), namely,

$$\text{IR}_p = 4n^2\sqrt{\frac{2}{\pi}}\,\frac{1}{2^n n!}\,\frac{1}{\omega}\int_0^{\infty}[H_{n-1}(g)]^2 e^{-g^2}\,\text{d}g = \frac{8\sqrt{2}}{\omega}\,n. \tag{84}$$

Thus, similar to IR_x, here also IR_p of a particular oscillator state can be recovered from the wave function of adjacent lower state. Equation (84) tells that, progress of IR_p with n is again linear like its x-space counterpart, slope of the straight line in this case being $\frac{8\sqrt{2}}{\omega}$. It is inversely proportional to ω in accordance with the fact that, with increase in oscillation, localization as well as fluctuation predominates. At the special value of $\omega = \sqrt{2}$, IR_x, IR_p become equal ($8n$). Apart from that, like the total energy difference in a 1DQHO, $\Delta\text{E}_\text{n}(= \text{E}_{\text{n}+1} - \text{E}_\text{n})$, the difference of IR between two successive states also remains constant, i.e., $\Delta(\text{IR}_x) = \text{IR}_x(n+1) - \text{IR}_x(n) = 4\sqrt{2}\omega$, and $\Delta(\text{IR}_p) = \text{IR}_p(n+1) - \text{IR}_p(n) = \frac{8\sqrt{2}}{\omega}$.

4.3.2. Isotropic 3D Quantum Harmonic Oscillator (3DQHO)

We start from the normalized r-space wave function given as,

$$\psi_{n_r,l}(r) = \sqrt{\frac{2\,\omega^{l+\frac{3}{2}}\,n_r!}{\Gamma(n_r + l + \frac{3}{2})}}\; r^l\, e^{-\frac{\omega r^2}{2}}\, L_{n_r}^{l+\frac{1}{2}}(\omega r^2). \tag{85}$$

Table 13. $E_{\mathbf{r}}, E_{\mathbf{p}}, E_t$ **for** $1s, 2s, 1p, 2p, 1d, 2d$ **states in 3DCHO at eight selected** r_c **[181]**

r_c	$E_{\mathbf{r}}$	$E_{\mathbf{p}}$	E_t	r_c	$E_{\mathbf{r}}$	$E_{\mathbf{p}}$	E_t
	$1s$				$2s$		
0.1	672.0719164	0.00000398	0.0026791	0.1	1453.1909702895	0.00000057	0.00082825
0.2	84.01080088	0.00003189	0.0026791	0.2	181.6485572148	0.00000455	0.00082825
0.5	5.3814002356	0.00049808	0.00268041	0.5	11.6246913489	0.00007124	0.00082821
1.0	0.6818097823	0.00396012	0.00270005	1.0	1.4515093698	0.00057014	0.00082757
5.0	0.0634936361	0.06349340	0.00403142	5.0	0.0406758398	0.04064829	0.00165340
8.0	0.0634936347	0.06349363	0.00403144	8.0	0.040675749	0.04067560	0.00165430
∞	0.0634936347	0.06349363	0.00403144	∞	0.0406756097	0.04067560	0.00165450
	$1p$				$2p$		
0.1	803.22700816	0.00000227	0.00182938	0.1	1454.974575234	0.00000058	0.00085788
0.2	100.40423515	0.00001822	0.00182940	0.2	181.871793182	0.00000471	0.00085788
0.5	6.4281046025	0.00028463	0.00182965	0.5	11.6397197874	0.00007369	0.00085777
1.0	0.8078468658	0.00226961	0.00183350	1.0	1.454810759	0.00058839	0.00085600
5.0	0.0476202385	0.04761162	0.00226727	5.0	0.0324128865	0.03237583	0.00104939
8.0	0.047620224	0.04762022	0.00226768	8.0	0.032411515	0.03241151	0.00105050
∞	0.047620224	0.04762022	0.00226768	∞	0.032411515	0.03241151	0.00105050
	$1d$				$2d$		
0.1	851.25726418	0.00000137	0.00117301	0.1	1368.86082394	0.00000047	0.00064967
0.2	106.40757650	0.00001102	0.00117301	0.2	171.107627763	0.00000379	0.00064967
0.5	6.8111728555	0.00017222	0.00117305	0.5	10.9509523492	0.00005931	0.00064960
1.0	0.8535077417	0.00137508	0.00117364	1.0	1.3689869817	0.00047372	0.00064852
5.0	0.0357152613	0.03571436	0.00127554	5.0	0.0249812774	0.02492595	0.00062268
8.0	0.0357151695	0.03571516	0.00127557	8.0	0.0249755634	0.02497556	0.00062377
∞	0.0357151695	0.03571516	0.00127557	∞	0.0249755634	0.02497556	0.00062377

In the above, $L_n^{\alpha}(x)$ represents the associated Laguerre polynomial. Now, using Eq. (85) and substituting $\omega r^2 = u$, first part of Eq. (76) yields,

$$\mathrm{IR}_{\mathbf{r}} = \frac{16\,\omega\, n_r!}{\Gamma(n_r+l+\frac{3}{2})} \int_0^{\infty} u^{l+\frac{3}{2}} e^{-u} \left[L_{n_r}^{l+\frac{1}{2}}(u)\right]^2 \left(\frac{\psi'_{n_r,l}(u)}{\psi_{n_r,l}(u)} - \frac{\psi'_{n_{r1},l}(u)}{\psi_{n_{r1},l}(u)}\right)^2 du. \tag{86}$$

Now, using the recurrence relation $\frac{d}{du}L_{n_r}^{l+\frac{1}{2}}(u) = -L_{n_r-1}^{l+\frac{3}{2}}(u)$, the ratios of wave functions given in the parentheses may be simplified as,

$$\frac{\psi'_{n_r,l}(u)}{\psi_{n_r,l}(u)} = \frac{l}{2u} - \frac{1}{2} - \frac{L_{n_r-1}^{l+\frac{3}{2}}(u)}{L_{n_r}^{l+\frac{1}{2}}(u)}, \tag{87}$$

the right-hand side of which, for a node-less state becomes $\left(\frac{l}{2u} - \frac{1}{2}\right)$, because the ratio of polynomials vanishes. Next one may invoke the usual orthonormality relation,

$$\int_0^\infty u^k e^{-u} L_i^k(u) L_j^k(u) \mathrm{d}u = \frac{(i+k)!}{i!}\,\delta_{ij}. \tag{88}$$

to obtain the final form of $\mathrm{IR}_\mathbf{r}$ as below,

$$\mathrm{IR}_\mathbf{r} = \frac{16\,\omega\, n_r!}{\Gamma(n_r+l+\frac{3}{2})} \int_0^\infty u^{l+\frac{3}{2}} e^{-u} \left[L_{n_r-1}^{l+\frac{3}{2}}\right]^2 \mathrm{d}u = 16\omega\, n_r = 8\omega(n-l). \tag{89}$$

Equation (89) predicts that, $\mathrm{IR}_\mathbf{r}$ in a 3DQHO, like its 1D counterpart, is also a linear function of n; however in this occasion the slope is 8ω and intercept is negative. For a fixed l, the slope progresses and intercept decreases with ω respectively. Further, at a certain ω, the intercept declines with rise of l. In this scenario, the particle gets more and more localized with progress of ω. The relative fluctuation with respect to reference state advances with ω. Dependence of $\mathrm{IR}_\mathbf{r}$ on ω is reminiscent to that of r-space Fisher information, $\mathrm{I}_\mathbf{r}$ [128]—both quantities escalate with ω.

Analogously, in p-space the normalized wave function can be expressed as [182],

$$\psi_{n_r,l}(p) = \sqrt{\frac{2\,n_r!}{\Gamma(n_r-l+\frac{3}{2})\ \omega^{l+\frac{3}{2}}}}\ p^l\, e^{-\frac{p^2}{2\omega}}\, L_{n_r}^{l+\frac{1}{2}}\left(\frac{p^2}{\omega}\right). \tag{90}$$

Substituting $\frac{p^2}{\omega} = \chi$, and going through some simple algebraic steps, one can derive,

$$\mathrm{IR}_\mathbf{p} = \frac{16\,n_r!}{\Gamma(n_r+l+\frac{3}{2})\,\omega} \int_0^\infty \chi^{l+\frac{3}{2}} e^{-\chi} \left[L_{n_r}^{l+\frac{1}{2}}(\chi)\right]^2 \left(\frac{\psi'_{n_r,l}(\chi)}{\psi_{n_r,l}(\chi)} - \frac{\psi'_{n_{r_1},l}(\chi)}{\psi_{n_{r_1},l}(\chi)}\right)^2 \mathrm{d}\chi. \tag{91}$$

Applying similar arguments as discussed earlier for $\mathrm{IR}_{\mathbf{r}}$ produces,

$$\frac{\psi'_{n_r,l}(\chi)}{\psi_{n_r,l}(\chi)} - \frac{\psi'_{n_{r1},l}(\chi)}{\psi_{n_{r1},l}(\chi)} = -\frac{L_{n_r-1}^{l+\frac{3}{2}}(\chi)}{L_{n_r}^{l+\frac{1}{2}}(\chi)}. \tag{92}$$

which upon applying in Eq. (91), leads to the following final expression, namely,

$$\mathrm{IR}_{\mathbf{p}} = 16\left[\frac{n_r!}{\Gamma(n_r+l+\frac{3}{2})\,\omega}\right]\int_0^\infty \chi^{l+\frac{3}{2}}e^{-\chi}\left[L_{n_r-1}^{l+\frac{3}{2}}\right]^2 \mathrm{d}\chi = \frac{16}{\omega}n_r = \frac{8}{\omega}(n-l). \tag{93}$$

Equation (93) implies that, $\mathrm{IR}_{\mathbf{p}}$, like $\mathrm{IR}_{\mathbf{r}}$, also linearly changes with n; the slope and intercept are $\frac{16}{\omega}$ and $-\frac{16l}{\omega}$ respectively. An increase in ω facilitates localization and fluctuation too. $\mathrm{IR}_{\mathbf{r}}$ progresses with ω, while $\mathrm{IR}_{\mathbf{p}}$ decreases, signifying higher fluctuation at larger n_r. Once again $\mathrm{IR}_{\mathbf{r}}$, $\mathrm{IR}_{\mathbf{p}}$ of a given n_r, l-state may be calculated from (n_r-1), $(l+1)$-state wave functions; this holds true in both spaces. We close the discussion by pointing out that, in parallel to 1D case, here also both $\Delta(\mathrm{IR}_{\mathbf{r}})$, $\Delta(\mathrm{IR}_{\mathbf{p}})$, at a given l, depend only on ω and remain unaltered with respect to n. They are expressed as in the following,

$$\begin{aligned}\Delta(\mathrm{IR}_{\mathbf{r}}) &= \mathrm{IR}_{\mathbf{r}}(n+1,\ l) - \mathrm{IR}_{\mathbf{r}}(n,\ l) = 8\omega \\ \Delta(\mathrm{IR}_{\mathbf{p}}) &= \mathrm{IR}_{\mathbf{p}}(n+1,\ l) - \mathrm{IR}_{\mathbf{p}}(n,\ l) = \frac{8}{\omega}.\end{aligned} \tag{94}$$

4.4. A Virial-Like Theorem

Recently a Virial-like equation has been derived by the authors for confined quantum systems [183]. It has the form,

$$\begin{aligned}&\langle \hat{T}^2\rangle_n - \langle \hat{T}\rangle_n^2 = \langle \hat{V}^2\rangle_n - \langle \hat{V}\rangle_n^2 \\ &\quad(\Delta \hat{T}_n)^2 = \langle \hat{V}\rangle_n\langle \hat{T}\rangle_n - \langle \hat{T}\hat{V}\rangle_n = (\Delta \hat{V}_n)^2 = \langle \hat{T}\rangle_n\langle \hat{V}\rangle_n - \langle \hat{V}\hat{T}\rangle_n.\end{aligned} \tag{95}$$

This relation interprets that, the magnitude of error incurred in $\langle \hat{T}\rangle_n$ and $\langle \hat{V}\rangle_n$ are identical. Now, one can easily conclude that, $\mathcal{E}_n$ is a sum of two average quantities but still provides exact result. It is due to the cancellation of errors between $\langle \hat{T}\rangle_n$ and $\langle \hat{V}\rangle_n$. In a confinement condition, validity of Eq. (95) can be verified by deriving the expressions of $\langle \hat{T}\hat{V}\rangle_{n,\ell}$, $\langle \hat{V}\hat{T}\rangle_{n,\ell}$, $\langle \hat{V}^2\rangle_{n,\ell}$ and $\langle \hat{V}\rangle_{n,\ell}$ (other integrals remain unchanged).

Table 14. $\mathcal{E}_n$, $(\Delta V_n)^2$, $(\Delta T_n)^2$, $\langle T\rangle_n\langle V\rangle_n - \langle TV\rangle_n$, $\langle T\rangle_n\langle V\rangle_n - \langle VT\rangle_n$ **values for** $n = 0, 1$ **states in 1DCHO at four** $(0.1, 0.5, 1, \infty)$ **values of** x_c. **See text for detail**

n	Property	$x_c = 0.1$	$x_c = 0.5$	$x_c = 1$	$x_c = \infty$
	$\mathcal{E}_0^\dagger$	123.3707084678	4.9511293232	1.2984598320	0.4999999999
	$(\Delta V_0)^2$	0.000000600468	0.0003747558	0.0058688193	0.1299999999
0	$(\Delta T_0)^2$	0.000000600466	0.0003747558	0.0058688193	0.1299999999
	$\langle T\rangle_0\langle V\rangle_0 - \langle TV\rangle_0$	0.000000600466	0.0003747558	0.0058688193	0.1299999999
	$\langle T\rangle_0\langle V\rangle_0 - \langle VT\rangle_0$	0.000000600466	0.0003747558	0.0058688193	0.1299999999
	$\mathcal{E}_1^\ddagger$	493.481633417	19.7745341792	5.0755820152	1.4999999999
	$(\Delta V_1)^2$	0.00000085445	0.00053374630	0.0084865378	0.3749999999
1	$(\Delta T_1)^2$	0.00000085434	0.00053374630	0.0084865378	0.3749999999
	$\langle T\rangle_1\langle V\rangle_1 - \langle TV\rangle_1$	0.00000085434	0.00053374630	0.0084865378	0.3749999999
	$\langle T\rangle_1\langle V\rangle_1 - \langle VT\rangle_1$	0.00000085434	0.00053374630	0.0084865378	0.3749999999

†Literature results [34] of $\mathcal{E}_0$ for $x_c = 0.1, 0.5, 1, \infty$ are: 123.37070846785, 4.9511293232541, 1.2984598320321, 0.5 respectively.
‡Literature results [34] of $\mathcal{E}_1$ for $x_c = 0.1, 0.5, 1, \infty$ are: 493.48163341761, 19.774534179208, 5.0755820152268, 1.5 respectively.

4.4.1. Symmetrically Confined Harmonic Oscillator (SCHO)

In a 1DCHO case, the desired expectation values will take following form:

$$\langle \hat{T}\hat{V}\rangle_n = \langle \hat{T}v(x)\rangle_n + \langle \hat{T}v_c\rangle_n = \langle \hat{T}v(x)\rangle_n. \tag{96}$$

One can utilize the property of Reimann integral to simplify,

$$\begin{aligned}\langle \hat{T}v_c\rangle_n = &\int_{-\infty}^{-x_c} \psi_n^*(x)\hat{T}v_c\psi_n(x)\mathrm{d}x + \\ &\int_{-x_c}^{x_c} \psi_n^*(x)\hat{T}v_c\psi_n(x)\mathrm{d}x + \int_{x_c}^{\infty} \psi_n^*(x)\hat{T}v_c\psi_n(x)\mathrm{d}x = 0\end{aligned} \tag{97}$$

The first and third integrals reduce to *zero* because $\psi_n(x) = 0$ when $x \geq |x_c|$, whereas the second integral leads to *zero* as, $v_c = 0$ inside the box. Similarly,

$$\langle \hat{V}\hat{T}\rangle_n = \langle v(x)\hat{T}\rangle_n + \langle v_c\hat{T}\rangle_n = \langle v(x)\hat{T}\rangle_n \tag{98}$$

$$\langle \hat{V}^2 \rangle_n = \langle (v(x)^2 \rangle_n + \langle v(x) v_c \rangle_n + \langle v_c v(x) \rangle_n + \langle v_c^2 \rangle_n = \langle (v(x))^2 \rangle_n \quad (99)$$

$$\langle \hat{V} \rangle_n = \langle v(x) \rangle_n + \langle v_c \rangle_n = \langle v(x) \rangle_n. \quad (100)$$

Thus, for a 1DCHO, with the help of above equations, Eq. (95) may be recast as,

$$\left(\Delta \hat{T}_n\right)^2 = \left(\Delta \hat{V}_n\right)^2 = \langle \hat{T} \rangle_n \langle v(x) \rangle_n - \langle v(x) \hat{T} \rangle_n = \langle \hat{T} \rangle_n \langle v(x) \rangle_n - \langle \hat{T} v(x) \rangle_n. \quad (101)$$

Thus it is clear from Eq. (101) that, v_c does not contribute to the desired expectation values. Hence the only difference between free and enclosed system is that, in latter case, the boundary has been reduced to a finite region from infinity. Representative numerical values of $\mathcal{E}_n$, $(\Delta \hat{T}_n)^2$, $(\Delta \hat{V}_n)^2$, $\langle T \rangle_n \langle V \rangle_n - \langle TV \rangle_n$ and $\langle T \rangle_n \langle V \rangle_n - \langle VT \rangle_n$ are produced in Table 14 for $n = 0,\ 1$ states of 1DCHO at four selected x_c values, namely $0.1,\ 0.5,\ 1,\ \infty$. In all these four x_c, $\mathcal{E}_0$ and $\mathcal{E}_1$ remain in excellent agreement with available literature results as compared in [34], and hence not repeated here. However, no direct reference could be found for expectation values to compare. It is easily noticed that, in both confined and free (last column) scenario, Eq. (95) is obeyed, as all expectation values offer equal values, which affirms the applicability of our newly designed theorem in case of 1DCHO. Additionally, with increase in x_c, both $\left(\Delta \hat{T}\right)^2$, $\left(\Delta \hat{V}\right)^2$ increase, which presumably occurs as the wave function delocalizes with x_c. Consequently, the difference between mean square and average values of $\hat{T}$, $\hat{V}$ tends to rise. For more details, please consult [183].

4.4.2. 3D Confined Harmonic Oscillator (3DCHO)

The relevant expectation values for 3DCHO will now take following forms,

$$\langle \hat{T} \hat{V} \rangle_{n_r,\ell} = \langle \hat{T} v(r) \rangle_{n_r,\ell} + \langle \hat{T} v_c(r) \rangle_{n_r,\ell} = \langle \hat{T} v(r) \rangle_{n_r,\ell}. \quad (102)$$

This happens because $\langle \hat{T} v_c(r) \rangle_{n_r,\ell} = 0$, which, in turn occurs, since wave function disappears for $r \geq r_c$. A similar argument ($\langle v_c(r) \hat{T} \rangle_{n_r,\ell} = 0$) leads to the following relation,

$$\langle \hat{V} \hat{T} \rangle_{n_r,\ell} = \langle v(r) \hat{T} \rangle_{n_r,\ell} + \langle v_c(r) \hat{T} \rangle_{n_r,\ell} = \langle v(r) \hat{T} \rangle_{n_r,\ell} \quad (103)$$

Then since $\langle v(r) v_c(r) \rangle_{n_r,\ell} = \langle v_c(r) v(r) \rangle_{n_r,\ell} = \langle v_c(r)^2 \rangle_{n_r,\ell} = 0$, we can write,

$$\begin{aligned}\langle \hat{V}^2 \rangle_{n_r,\ell} &= \langle v(r)^2 \rangle_{n_r,\ell} + \langle v(r) v_c(r) \rangle_{n_r,\ell} + \langle v_c(r) v(r) \rangle_{n_r,\ell} + \langle v_c(r)^2 \rangle_{n_r,\ell} \\ &= \langle v(r)^2 \rangle_{n_r,\ell}.\end{aligned} \quad (104)$$

Table 15. $\mathcal{E}_{n_r,\ell}$ **,**$(\Delta V_{n_r,\ell})^2$ **,**$(\Delta T_{n_r,\ell})^2$ **,**$\langle T\rangle_{n_r,\ell}\langle V\rangle_{n_r,\ell} - \langle TV\rangle_{n_r,\ell}$ **,**$\langle T\rangle_{n_r,\ell}\langle V\rangle_{n_r,\ell} - \langle VT\rangle_{n_r,\ell}$ **for** $1s,\ 1p,\ 2s$ **states in 3DCHO at four specific** r_c**'s, such as** $0.1, 0.5, 1, \infty$

State	Property	$r_c = 0.1$	$r_c = 0.5$	$r_c = 1$	$r_c = \infty$
$1s$	$\mathcal{E}_{1,0}^{¶}$	493.481633459	19.774534179	5.0755820153	1.4999999
	$(\Delta V_{1,0})^2$	0.00000085434	0.0005337463	0.0084865378	0.3749999
	$(\Delta T_{1,0})^2$	0.00000085434	0.0005337463	0.0084865378	0.3749999
	$\langle T\rangle_{1,0}\langle V\rangle_{1,0} - \langle TV\rangle_{1,0}$	0.00000085434	0.0005337463	0.0084865378	0.3749999
	$\langle T\rangle_{1,0}\langle V\rangle_{1,0} - \langle VT\rangle_{1,0}$	0.00000085434	0.0005337463	0.0084865378	0.3749999
$1p$	$\mathcal{E}_{1,1}^{§}$	1009.53830080	40.428276496	10.282256939	2.4999999
	$(\Delta V_{1,1})^2$	0.0000008424	0.0005264224	0.0084064867	0.6249999
	$(\Delta T_{1,1})^2$	0.00000084238	0.0005264224	0.0084064867	0.6249999
	$\langle T\rangle_{1,1}\langle V\rangle_{1,1} - \langle TV\rangle_{1,1}$	0.00000084238	0.0005264224	0.0084064867	0.6249999
	$\langle T\rangle_{1,1}\langle V\rangle_{1,1} - \langle VT\rangle_{1,1}$	0.00000084238	0.0005264224	0.0084064867	0.6249999
$2s$	$\mathcal{E}_{2,0}^{\ddagger}$	1973.92248339	78.9969211469	19.899696502	3.4999999
	$(\Delta V_{2,0})^2$	0.00000182	0.00113739969	0.0181584455	1.6249999
	$(\Delta T_{2,0})^2$	0.00000182	0.00113739969	0.0181584455	1.6249999
	$\langle T\rangle_{2,0}\langle V\rangle_{2,0} - \langle TV\rangle_{2,0}$	0.00000182	0.00113739969	0.0181584455	1.6249999
	$\langle T\rangle_{2,0}\langle V\rangle_{2,0} - \langle VT\rangle_{2,0}$	0.00000182	0.00113739969	0.0181584455	1.6249999

¶Literature results [184] of $\mathcal{E}_{1,0}$ for $r_c = 0.1, 0.5, 1, \infty$ are: 493.48163346, 19.774534180, 5.0755820154, 1.5 respectively.

§Literature results [184] of $\mathcal{E}_{1,1}$ for $r_c = 0.1, 0.5, 1, \infty$ are: 1009.5383008, 40.428276496, 10.282256939, 2.5 respectively.

‡ Literature results [184] of $\mathcal{E}_{2,0}$ for $r_c = 0.1, 0.5, 1, \infty$ are: 1973.922483399, 78.996921147, 19.899696502, 3.5 respectively.

And finally, one can derive (since $\langle v_c(r)\rangle_{n_r,\ell} = 0$),

$$\langle\hat{V}\rangle_{n_r,\ell} = \langle v(r)\rangle_{n_r,\ell} + \langle v_c(r)\rangle_{n_r,\ell} = \langle v(r)\rangle_{n_r,\ell}. \tag{105}$$

Thus, for a 3DCHO, Eq. (95) can be recast in to,

$$\begin{aligned}\langle\hat{T}^2\rangle_{n_r,\ell} - \langle\hat{T}\rangle^2_{n_r,\ell} &= \langle\hat{V}^2\rangle_{n_r,\ell} - \langle\hat{V}\rangle^2_{n_r,\ell} \\ (\Delta\hat{T}_{n_r,\ell})^2 = (\Delta\hat{V}_{n_r,\ell})^2 &= \langle\hat{T}\rangle_{n_r,\ell}\langle v(r)\rangle_{n_r,\ell} - \langle v(r)\hat{T}\rangle_{n_r,\ell} \\ &= \langle\hat{T}\rangle_{n_r,\ell}\langle v(r)\rangle_{n_r,\ell} - \langle\hat{T}v(r)\rangle_{n_r,\ell}.\end{aligned}$$

This suggests that, similar to 1DCHO, here also the confining potential has no contribution on desired expectation values; only the boundary in confined system gets shifted to r_c, from ∞ of corresponding free counterpart. It clearly supports the validity of Eq. (95) in a 3DCHO. As an illustration, Table 15 imprints numerically calculated expectation values, for three low-lying ($1s$, $1p$, $2s$) states at four chosen values of confinement radius, i.e., $0.1,\ 0.5,\ 1,\ \infty$. This again establishes the efficiency of Eq. (95) for such potential in both confined and free system, as evident from identical values of these quantities at all r_c's–last column signifying the corresponding *unconstrained* system. Accurate energies are quoted from GPS results [184]. No literature is available for average values considered here. Like the 1D case, here also $(\Delta\hat{T}_{n_r,\ell})^2$, $(\Delta\hat{V}_{n_r,\ell})^2$ increase with r_c. A more detailed discussion may be found in [183].

Conclusion

Information-based uncertainty measures like I, S, R, E and several complexities in composite r, p spaces have been employed to understand the effect of confinement in 3DCHO. Before that, we have also discussed I, S, E and various complexities for SCHO and ACHO systems. Apart from that, relative Fisher information for 1DQHO and 3DQHO are reported here. A detailed study reveals that, an SCHO can be considered as an interim model between PIB and 1DQHO. Similarly, a 3DCHO may be treated as a two-mode system; at $r_c \rightarrow 0$ it behaves as PISB and at $r_c \rightarrow \infty$ it reduces to a 3DQHO.

It is imperative to mention that at $d_m = 0$, an ACHO reduces to an SCHO. An increase in d_m leads to localization of the system in real space. S_x, S_p can completely interpret the competing nature of ACHO potential. In contrast, I_x, I_p prove inefficient for such analysis. Further, E_x, E_p provide important knowledge which guides the reader to a better understanding of such systems.

In 3DCHO, at very low r_c, $R_{\mathbf{r}}^{\alpha}$, $S_{\mathbf{r}}$ reduce and $E_{\mathbf{r}}$ grows up as n_r progresses. This behavior clearly suggests that the effect of confinement accelerates with increase of number of nodes in the system. In future, it will be interesting to study the effect of penetrable boundary on such systems. Further, study of relative information for SCHO, ACHO and 3DCHO systems would be highly desirable.

Acknowledgment

The authors express their sincere gratitude to Prof. Nadya S. Columbus, President, Nova Science Publishers, NY, USA, for the kind invitation to present our work in this exciting area. We are extremely thankful to the Editor, Prof. Yilun Shang for giving us the opportunity to contribute and also extending the submission deadline generously. AKR gratefully acknowledges financial support from SERB, Department of Science and Technology (DST), New Delhi (sanction no. EMR/2014/000838). NM thanks DST for financial support through NPDF grant (PDF/2016/000014CS). It is a pleasure to thank Ms. Sangita Majumdar for giving a careful reading of the manuscript.

REFERENCES

[1] Michels, A.; De Boer, J.; Bijl, A. (1937). Remarks concerning molecular interaction and their influence on the polarizability. *Physica 4*, 981.

[2] Jaskólski, W. (1996). Confined many-electron systems. *Phys. Rep. 271*, 1.

[3] Dolmatov, V. K.; Baltenkov, A. S.; Connerade, J.-P; Mason, S. T. (2004). Structure and photoionization of confined atoms. *Radiat. Phys. Chem. 70*, 417.

[4] Sabin, J., Brändas, E.; and Cruz, S. (Ed.) (2009). *Adv. Quant. Chem.*, Volume 57 & 58. New York: Academic Press.

[5] Sen, K. D. (Ed.) (2014). *Electronic Structure of Quantum Confined Atoms and Molecules*. Switzerland: Springer International Publishing.

[6] Ley-Koo, E. (2018). Recent progress in confined atoms and molecules: superintegrability and symmetry breakings. *Revista Mexicana de Física 64*, 326.

[7] Dolmatov, V. K. (2009). Photoionization of atoms encaged in spherical fullerenes. *Adv. Quant. Chem. 58*, 13.

[8] Charkin, O. P.; Klimenko, N. M.; Charkin, D. O. (2009). DFT study of molecules confined inside fullerene and fullerene-like cages. *Adv. Quant. Chem. 58*, 69.

[9] Etindele, A. J.; Maezono, R.; Melingui-Melono, R. L.; Motapon, O. (2017). Influence of endohedral confinement of atoms on structural and dynamical properties of the C_{60} fullerene. *Chem. Phys. Lett. 685*, 395.

[10] Auluck, F. C. (1941a). Energy levels of an artificially bounded linear oscillator. *Proc. Nat. Inst. Sci. India 7*, 133.

[11] Auluck, F. C. (1941b). The Artificially Bounded Relativistic Linear Oscillator. *Proc. Nat. Inst. Sci. India 7*, 383.

[12] Chandrasekhar, S. (1943). Dynamical friction. II. the rate of escape of stars from clusters and the evidence for the operation of dynamical friction. *Astrophys. J. 97*, 263.

[13] Dingle, R. B. (1952a). Some magnetic properties of metals III. Diamagnetic resonance. *Proc. Royal Soc. London A 212*, 38.

[14] Dingle, R. B. (1952b). Some magnetic properties of metals IV. Properties of small systems of electrons. *Proc. Royal Soc. London A 212*, 47.

[15] Grinberg, M.; Jaskólski, W.; Koepke, Cz.; Planelles, J.; Janowicz, M. (1994). Spectroscopic manifestation of a confinement-type lattice anharmonicity. *Phys. Rev. B 50*, 6504(R).

[16] Auluck, F. C.; Kothari, D. S. (1945). The quantum mechanics of a bounded linear harmonic oscillator. *Proc. Cambridge Phil. Soc. 41*, 175.

[17] Baijal, J. S.; Singh, K. K. (1955). The energy levels and transition probabilities for a bounded linear harmonic oscillator. *Prog. Theor. Phys. (Kyoto) 14*, 214.

[18] Hull, T. E.; Julius, R. S. (1956). Enclosed quantum mechanical systems. *Can. J. Phys. 34*, 914.

[19] Dean, P. (1966). The constrained quantum mechanical harmonic oscillator. *Proc. Cambridge Phil. Soc. 62*, 277.

[20] Vawter, R. (1968). Effects of finite boundaries in a one-dimensional harmonic oscillator. *Phys. Rev. 174*, 749.

[21] Vawter, R. (1973). Energy eigenvalues of a bounded centrally located harmonic oscillator. *J. Math. Phys. 14*, 1864.

[22] Consortini, A.; Frieden, B. R. (1976). Quantum-mechanical solution for the simple harmonic oscillator in a box. *Nuovo Cimento B 35*, 153.

[23] Aguilera-Navarro, V.C.; Ley-Koo, E.; Zimerman, A. H. (1980). Perturbative, asymptotic and Padé-approximant solutions for harmonic and inverted oscillators in a box. *J. Phys. A 13*, 3585.

[24] Fernández, F. M.; Castro, E. A. (1981a). Hyper-virial analysis of enclosed quantum mechanical systems. I. Dirichlet boundary conditions. *Int. J. Quant. Chem. 19*, 521.

[25] Fernández, F. M.; Castro, E. A. (1981b). Hyper-virial analysis of enclosed quantum mechanical systems. III. Unsymmetrical boundary conditions. *Int. J. Quant. Chem. 20*, 623.

[26] Arteca, G. A.; Maluendes, S. A.; Fernández, F. M.; Castro, E. A. (1983). Discussion of several analytical expressions for the eigenvalues of the bounded harmonic oscillator and hydrogen atom. *Int. J. Quant. Chem. 24*, 169.

[27] Taşeli, H. (1993). Accurate computation of the energy spectrum for potentials with multiminima. *Int. J. Quant. Chem. 46*, 319.

[28] Vargas, R.; Garza, J.; Vela, A.; (1996). Strongly convergent method to solve one-dimensional quantum problems. *Phys. Rev. E 53*, 1954.

[29] Sinha, A.; Roychoudhury, R. (1999). WKB and MAF quantization rules for spatially confined quantum mechanical systems. *J. Math. Chem. 73*, 497.

[30] Sinha, A. (2000). SWKB formalism for confined quantum systems. *Int. J. Quant. Chem. 79*, 267.

[31] Montgomery Jr., H. E.; Campoy, G.; Aquino, N. (2010). The confined N-dimensional harmonic oscillator revisited. *Phys. Scr. 81*, 045010.

[32] Campoy, G.; Aquino, N.; Granados, V. N. (2002). Energy eigenvalues and Einstein coefficients for the one-dimensional confined harmonic oscillators. *J. Phys. A 35*, 4903.

[33] Amore, P.; Fernández, F. M. (2010). One-dimensional oscillator in a box. *Eur. J. Phys. 31*, 69.

[34] Roy, A. K. (2015). Quantum confinement in 1D systems through an imaginary-time evolution method. *Mod. Phys. Lett. A 37*, 1550176.

[35] Aquino, N.; Cruz, E. (2017). The 1-dimensional confined harmonic oscillator revisited. *Revista Mexicana de Física 63*, 580.

[36] Aquino, N.; Castaño, E.; Campoy, G.; Granados, V. (2001). Einstein coefficients and dipole moment for the asymmetrically confined harmonic oscillator. *Eur. J. Phys. 22*, 645.

[37] Fernández, F. M.; Castro, E. A. (1981). Hyper-virial treatment of multidimensional isotropic bounded oscillators. *Phys. Rev. A 24*, 2883.

[38] Aguilera-Navarro, V. C.; Gomes, J. F.; Zimerman, A. H.; Ley-Koo, E. (1983). On the radius of convergence of Rayleigh-Schrödinger perturbative solutions for quantum oscillators in circular and spherical boxes. *J. Phys. A 16*, 2943.

[39] Marin, J. L.; Cruz, S. A. (1991). On the use of direct variational methods to study confined quantum systems. *Am. J. Phys. 59*, 931.

[40] Dutt, R.; Mukherjee, A.; Varshni, Y. P. (1995). Supersymmetric semiclassical approach to confined quantum problems. *Phys. Rev. A 52*, 1750.

[41] Taşeli, H.; Zafer, A. (1997a). A Fourier-Bessel expansion for solving radial Schrödinger equation in two dimensions. *Int. J. Quant. Chem. 61*, 759.

[42] Taşeli, H.; Zafer, A. (1997b). Bessel basis with applications: N-dimensional isotropic polynomial oscillators. *Int. J. Quant. Chem. 63*, 936.

[43] Aquino, N. (1997). The isotropic bounded oscillators. *J. Phys. A 30*, 2403.

[44] Sinha, A. (2003). Three-dimensional confinement: WKB revisited. *J. Math. Chem. 34*, 201.

[45] Filho, E. D.; Ricotta, R. M. (2003). Supersymmetric variational energies of 3D confined potentials. *Phys. Lett. A 320*, 95.

[46] Sen, K. D.; Roy, A. K. (2006). Studies on the 3D confined potentials using generalized pseudospectral approach. *Phys. Lett. A 357*, 112.

[47] Roy, A. K. (2014). Confinement in 3D polynomial oscillators through a generalized pseudospectral method. *Mod. Phys. Lett. A 29*, 1450104.

[48] Stevanović, Lj.; Sen, K. D. (2008a). A study of the confined 2D isotropic harmonic oscillator in terms of the annihilation and creation operators and the infinitesimal operators of the SU(2) group. *J. Phys. A 41*, 265203.

[49] Stevanović, Lj.; Sen, K. D. (2008b). Eigenspectrum properties of the confined 3D harmonic oscillator. *J. Phys. A 41*, 225002.

[50] Serrano, F. A.; Dong, S.-H. (2013). Proper quantization rule approach to three-dimensional quantum dots. *Int. J. Quant. Chem. 113*, 2282.

[51] Al-Jaber, S. M. (2008). A confined N-dimensional harmonic oscillator. *Int. J. Theor. Phys. 47*, 1853.

[52] Montgomery, Jr., H. E.; Aquino, N. A.; Sen, K. D. (2007). Degeneracy of confined D-dimensional harmonic oscillator. *Int. J. Quant. Chem. 107*, 798.

[53] Poland, D. (2000). Maximum-entropy calculation of energy distributions. *J. Chem. Phys. 112*, 6554.

[54] Singer, A. (2004). Maximum entropy formulation of the Kirkwood superposition approximation. *J. Chem. Phys. 121*, 3657.

[55] Antoniazzi, A.; Fanelli, D.; Barré, J.; Chavanis, P.-H.; Dauxois, T.; Ruffo, S. (2007). Maximum entropy principle explains quasistationary states in systems with long-range interactions: The example of the Hamiltonian mean-field model. *Phys. Rev. E 75*, 011112.

[56] Bialyniki-Birula, I.; Mycielski, J. (1975). Uncertainty relation for information entropy in wave mechanics. *Commun. Math. Phys. 44*, 129.

[57] Shannon, C. E. (1951). Prediction and entropy of printed english. *Bell System Technical J. 30*, 50.

[58] Bialyniki-Birula, I. (2006). Formulation of uncertainty relations in terms of the Rényi entropies. *Phys. Rev. A 74*, 052101.

[59] Duan, L.-M.; Giedke. G.; Cirac, J. I.; Zoller, P. (2000). Inseparability criterion for continuous variable systems. *Phys. Rev. Lett. 84*, 2722.

[60] Simon, R. (2000). Peres-Horodecki separability criterion for continuous variable systems. *Phys. Rev. Lett. 84*, 2726.

[61] Roy, A. K. (2004a). Calculation of the spiked harmonic oscillators through a generalized pseudospectral method. *Phys. Lett. A 321*, 231.

[62] Roy, A. K. (2004b). Calculation of the bound states of power-law and logarithmic potentials through a generalized pseudospectral method. *J. Phys. G 30*, 269.

[63] Roy, A. K. (2004c). Studies on the hollow states of atomic lithium using a density functional approach. *J. Phys. B 37*, 4569.

[64] Roy, A. K. (2005a). The generalized pseudospectral approach to the bound states of Húlthen and Yukawa potential. *Pramana-J. Phys. 65*, 1.

[65] Roy, A. K. (2005b). Studies on some singular potentials in quantum mechanics. *Int. J. Quant. Chem. 104*, 861.

[66] Roy, A. K. (2005c). Density functional studies on the hollow resonances in Li-isoelectronic sequence (Z=4-10). *J. Phys. B 38*, 1591.

[67] Roy, A. K.; Jalbout, A. F. (2007). Ground and excited states of Li^-, Be^- through a density-based approach. *Chem. Phys. Lett. 445*, 355.

[68] Roy, A. K.; Jalbout, A. F.; Proynov, E. I. (2008a). Accurate calculation of the bound states of hellmann potential. *J. Math. Chem. 44*, 260.

[69] Roy, A. K.; Jalbout, A. F.; Proynov, E. I. (2008b). Bound state spectra of the 3D rational potential. *Int. J. Quant. Chem. 108*, 827.

[70] Roy, A. K.; Jalbout, A. F. (2008). Bound states of the generalized spiked harmonic oscillator. *J. Mol. Struct. (Theochem) 853*, 27.

[71] Roy, A. K. (2013a). Studies on some exponential-screened Coulomb potentials. *Int. J. Quant. Chem. 113*, 1503.

[72] Roy, A. K. (2013b). Accurate ro-vibrational spectroscopy of diatomic molecules in a Morse oscillator potential. *Results in Physics 3*, 103.

[73] Roy, A. K. (2014a). Studies on the bound-state spectrum of Hyperbolic potential. *Few-Body Systems 55*, 143.

[74] Roy, A. K. (2014b). Ro-vibrational studies of diatomic molecules in a shifted Deng-Fan oscillator potential. *Int. J. Quant. Chem. 114*, 383.

[75] Roy, A. K. (2014c). Studies on bound-state spectra of Manning-Rosen potential. *Mod. Phys. Lett. A 29*, 1450042.

[76] Roy, A. K. (2014d). Ro-vibrational spectroscopy of molecules represented by Tietz-Hua oscillator potential. *J. Math. Chem 52*, 1405.

[77] Roy, A. K. (2015). Spherical confinement of Coulombic systems inside an impenetrable box: H atom and the Húlthen potential. *Int. J. Quant. Chem. 115*, 937.

[78] Roy, A. K. (2016). Critical parameters and spherical confinement of H atoms in screened Coulomb potential. *Int. J. Quant. Chem. 116*, 953.

[79] Abramowitz, M.; Stegun, I. (1964). *Handbook of Mathematical Functions*. New York: Dover.

[80] Anderson; J. B. (1975). A randon-walk simulation of the Schrödinger equation: H_3^+. *J. Chem. Phys. 63*, 1499.

[81] Kosloff, R.; Tal-Ezer, H. (1986). A direct relaxation method for calculating eigenfunctions and eigenvalues of the Schrödinger equation on a grid. *Chem. Phys. Lett. 127*, 223.

[82] Lehtovaara, L.; Toivanen, J.; Eloranta, J. (2007). Solution of time-dependent Schrödinger equation by the imaginary-time propagation method. *J. Comput. Phys. 221*, 148.

[83] Chin, S. A.; Janecek, S.; Krotscheck, S. (2009). Any order imaginary time propagation method for solving the Schrödinger equation. *Chem. Phys. Lett. 470*, 342.

[84] Sudiarta, I. W.; Wallace Geldart, D. J. (2009). The finite difference time domain method for computing the single-particle density matrix. *J. Phys. A 42*, 285002.

[85] Strickland, M.; Yager-Elorriaga, D. (2010). A parallel algorithm for solving the 3d Schrödinger equation. *J. Comput. Phys. 229*, 6015.

[86] Luukko, P. J. J.; Räsänen, E. (2013). Imaginary time propagation code for large-scale two-dimensional eigenvalue problems in magnetic fields. *Comput. Phys. Commun. 184*, 769.

[87] Roy, A. K.; Dey, B. K.; Deb, B. M. (1999). Direct calculation of ground-state electronic densities and properties of noble gas atoms through a single time-dependent hydrodynamical equation. *Chem. Phys. Lett. 308*, 523.

[88] Roy, A. K.; Chu, S. I. (2002). Quantum fluid dynamics approach for electronic structure calculation: Application to the study of ground-state properties of rare gas atoms. *J. Phys. B 35*, 2075.

[89] Roy A. K.; Gupta N.; Deb B. M. (2002). Time-dependent quantum-mechanical calculation of ground and excited states of anharmonic and double-well oscillators. *Phys. Rev. A 65*, 012109.

[90] Gupta N.; Roy A. K.; Deb, B. M. (2002). One-dimensional multiple-well oscillators: A time-dependent quantum mechanical approach. *Pramana J. Phys. 59*, 575.

[91] Wadehra, A.; Roy, A. K.; Deb, B. M. (2003). Ground and excited states of one-dimensional self-interacting nonlinear oscillators through time-dependent quantum mechanics. *Int. J. Quant. Chem. 91*, 597.

[92] Roy, A. K.; Thakkar, A. J.; Deb, B. M. (2005). Low-lying states of two-dimensional double-well potentials. *J. Phys. A 38*, 2189.

[93] Roy, A. K. (2014). Ground and excited states of spherically symmetric potentials through an imaginary-time evolution method: application to spiked harmonic oscillators. *J. Math. Chem 52*, 2645.

[94] Hammond, B. L.; Lester Jr., W. A.; Reynolds, P. J. (1994). *Monte Carlo Methods in Ab Initio Quantum Chemistry*. Singapore: World Scientific.

[95] Roy, A. K.; Chu, S. I. (2002a). Density functional calculations on singly- and doubly-excited rydberg states of many-electron systems. *Phys. Rev. A 65*, 052508.

[96] Roy, A. K.; Chu, S. I. (2002b). Quantum fluid dynamics approach for strong field processes: Aapplication to the study of multiphoton ionization and high-order harmonic generation of He and Ne in intense laser fields. *Phys. Rev. A 65*, 043402.

[97] Mukherjee, N; Roy, A. K. (2018). Information-entropic measures in free and confined hydrogen atom. *Int. J. Quant. Chem. 118*, e25596.

[98] Cover, T. M.; Thomas, J. A. (2006). *Elements of Information Theory*. New York: John Wiley & Sons, Ltd.

[99] Nielsen, M. A.; Chuang, I. L. (2010). *Quantum Computation and Quantum Information*. Cambridge: Cambridge University Press.

[100] Ramírez, J. C.; Soriano, C.; Esquivel, R. O.; Sagar, R. P.; Hô, M.; Smith, V. H. (1997). Jaynes information entropy of small molecules: Numerical evidence of the Collins conjecture. *Phys. Rev. A 56*, 4477.

[101] Delle, L. S. (2015). Shannon entropy and many-electron correlations: Theoretical concepts, numerical results, and Collins conjecture. *Int. J. Quant. Chem. 115*, 1396.

[102] Ghiringhelli, L. M.; Hamilton, I. P.; Delle, L. S (2010). Interacting electrons, spin statistics, and information theory. *J. Chem. Phys. 132*, 014106.

[103] Ghiringhelli, L. M.; Delle, L. S.; Mosna, R. A.; Hamilton, I. P. (2010). Information-theoretic approach to kinetic-energy functionals: the nearly uniform electron gas. *J. Math. Chem. 48*, 78.

[104] Nagy, Á. (2014). Fisher and Shannon information in orbital-free density functional theory. *Int. J. Quant. Chem. 115*, 1392.

[105] Alipour, M. (2015). Making a happy match between orbital-free density functional theory and information energy density. *Chem. Phys. Lett. 635*, 210.

[106] Delle, L. S. (2009a). On the scaling properties of the correlation term of the electron kinetic functional and its relation to the Shannon measure. *Euro Phys. Lett. 86*, 40004.

[107] Delle, L. S. (2009b). Erratum: On the scaling properties of the correlation term of the electron kinetic functional and its relation to the Shannon measure. *Euro Phys. Lett. 88*, 19901.

[108] Mohajeri, A.; Alipour, M. (2009). Information energy as an electron correlation measure in atomic and molecular systems. *Int. J. Quantum Inf. 07*, 801.

[109] Grassi, A. (2011). A relationship between atomic correlation energy of neutral atoms and generalized entropy. *Int. J. Quant. Chem. 111*, 2390.

[110] Flores-Gallegos, N. (2016). Informational energy as a measure of electron correlation. *Chem. Phys. Lett. 666*, 62.

[111] Alipour, M.; Badooei, Z. (2018). Toward electron correlation and electronic properties from the perspective of information functional theory. *J. Phys. Chem. A 122*, 6424.

[112] Barghathi, H.; Herdman, C.; Maestro, A. D. (2018). Rényi generalization of the accessible entanglement entropy. *Phys. Rev. Lett. 121*, 150501.

[113] Noorizadeh, S.; Shakerzadeh, E. (2010). Shannon entropy as a new measure of aromaticity, Shannon aromaticity. *Phys. Chem. Chem. Phys. 12*, 4742.

[114] Rényi, A. (1961). On measures of entropy and information. *Proc. Fourth Berkeley Symp. on Math. Statist. and Prob.*, 547.

[115] Rényi, A. (1970). *Probability Theory*. Amsterdam: North-Holland Pub. Company.

[116] Varga, I.; Pipek, J. (2003). Rényi entropies characterizing the shape and the extension of the phase space representation of quantum wave functions in disordered systems. *Phys. Rev. E 68*, 026202.

[117] Renner, R.; Gisin, N.; Kraus. B. (2005). Information-theoretic security proof for quantum-key-distribution protocols. *Phys. Rev. A 72*, 012332.

[118] Lévay, P.; Nagy, S.; Pipek, J. (2005). Elementary formula for entanglement entropies of fermionic systems. *Phys. Rev. A 72*, 022302.

[119] Verstraete, F.; Cirac, J. I. (2006). Matrix product states represent ground states faithfully. *Phys. Rev. B 73*, 094423.

[120] Bialas, A.; Czyz, W.; Zalewski, K. (2006). Moments of the Wigner function and Rényi entropies at freeze-out. *Phys. Rev. C 73*, 034912.

[121] Salcedo, L. L. (2009). Phase-space localization of antisymmetric functions. *J. Math. Phys. 50*, 012106.

[122] Nagy, Á.; Romera. E. (2009). Maximum Rényi entropy principle and the generalized Thomas-Fermi model. *Phys. Lett. A 373*, 844.

[123] McMinis, J.; Tubman, N. M. (2013). Rényi entropy of the interacting fermi liquid. *Phys. Rev. B 87*, 081108(R).

[124] Liu, S.-B.; Rong, C.-Y.; Wu, Z.-M.; Lu, T. (2015). Moments of the Wigner function and Rényi entropies at freeze-out. *Acta. Phys. Chim. Sin. 31*, 2057.

[125] Kim, I. (2018). Rényi-α entropies of quantum states in closed form: Gaussian states and a class of non-Gaussian states. *Phys. Rev. E 97*, 062141.

[126] Frieden, B. R. (2004). *Science from Fisher Information: A Unification*. Cambridge: Cambridge University Press.

[127] Sen, K. D. (Ed.) (2011). *Statistical Complexity: Applications in Electronic Structure*. Netherland: Springer International Publishing.

[128] Romera, E.; Sánchez-Moreno, P.; Dehesa, J. S. (2005). The Fisher information of single-particle systems with a central potential. *Chem. Phys. Lett. 414*, 468.

[129] Mukherjee, N.; Majumdar, S.; Roy, A. K. (2018). Fisher information in confined hydrogen-like ions. *Chem. Phys. Lett. 691*, 449.

[130] Mukherjee, N.; Roy, A. K. (2018). Fisher information in confined isotropic harmonic oscillator. *Int. J. Quant. Chem. 118*, e25727.

[131] Patil, S. H.; Sen, K. D.; Watson, N. A.; Montgomery Jr., H. E. (2007). Characteristic features of net information measures for constrained Coulomb potentials. *J. Phys. B 40*, 2147.

[132] Ghosal, A.; Mukherjee, N.; Roy, A. K. (2016). Information-entropic measures of a quantum harmonic oscillator in symmetric and asymmetric confinement within an impenetrable box. *Ann. Phys. (Berlin) 528*, 796.

[133] Onicescu, O. (1966). Energie informationnelle. *Comptes Rendus Hebdomadaires des Seances de lAcademie des Sciences Serie A 263*, 841.

[134] Pardo, L. (1986). Order-α weighted information energy. *Info. Sciences 40*, 155.

[135] Bhatia, P. (1997). On measures of information energy. *Info. Sciences 97*, 233.

[136] Noughabi, H. A.; Chahkandi, M. (2015). Informational energy and its application in testing normality. *Ann. Data Sci. 2*, 391.

[137] Alcoba, D. R.; Torre, A.; Lain, L.; Massaccesi, G. E.; Oña, O. B.; Ayers, P. W.; Van Raemdonck, M.; Bultinck, P.; Van Neck, D. (2016). Performance of Shannon-entropy compacted N-electron wave functions for configuration interaction methods. *Theor. Chem. Acc. 135*, 153.

[138] Zozor, S.; Portesi, M.; Vignat, C. (2008). Some extensions of the uncertainty principle. *Physica A 387*, 4800.

[139] Shiner, J. S.; Landsberg, P. T.; Davison, M. (1999). Simple measure for complexity. *Phys. Rev. E 59*, 1459.

[140] Catalán, R. G.; Garay, J.; López-Ruiz, R. (2002). Features of the extension of a statistical measure of complexity to continuous systems. *Phys. Rev. E 66*, 11102.

[141] Sánchez, J. R.; López-Ruiz, R. (2005). A method to discern complexity in two-dimensional patterns generated by coupled map lattices. *Physica A 355*, 633.

[142] Landsberg, P. T.; Shiner; J. S. (1998). Disorder and complexity in an ideal non-equilibrium fermi gas. *Phys. Lett. A 245*, 228.

[143] Romera, E.; Dehesa, J. S. (2004). The Fisher-Shannon information plane, an electron correlation tool. *J. Chem. Phys. 120*, 8906.

[144] Sen, K. D.; Antolín, J.; Angulo, J. C. (2007). Fisher-Shannon analysis of ionization processes and isoelectronic series. *Phys. Rev. A 76*, 032502.

[145] Sen, K. D.; Antolín, J.; Angulo, J. C. (2008). Fisher-Shannon plane and statistical complexity of atoms. *Phys. Rev. A 372*, 670.

[146] Antolín, J.; Angulo, J. C. (2009). Complexity analysis of ionization processes and isoelectronic series. *Int. J. Quant. Chem. 109*, 586.

[147] Calbet, X.; López-Ruiz, R. (2001). Tendency towards maximum complexity in a nonequilibrium isolated system. *Phys. Rev. E 63*, 066116.

[148] Martin, M. T.; Plastino, A.; Rosso, A. O. (2006). Generalized statistical complexity measures: Geometrical and analytical properties. *Physica A 369*, 439.

[149] Romera E., Nagy, Á. (2008). Rényi information of atoms. *Phys. Lett. A 372*, 4918.

[150] Angulo, J. C.; Antolín, J. (2008). Atomic complexity measures in position and momentum spaces. *J. Chem. Phys. 128*, 164109.

[151] Esquivel, R. O.; Angulo, J. C.; Antolín, J.; Dehesa, J. S.; López-Rosa, S.; Flores-Gallegos, N. (2010). Analysis of complexity measures and information planes of selected molecules in position and momentum spaces. *Phys. Chem. Chem. Phys. 12*, 7108.

[152] Welearegay, M. A.; Balawender, R.; Holas A. (2014). Information and complexity measures in molecular reactivity studies. *Phys. Chem. Chem. Phys. 16*, 14928.

[153] Esquivel, R. O.; Molina-Espíritu, M.; Angulo, J. C.; Antolín, J.; Flores-Gallegos, N.; Dehesa, J. S. (2011). Information-theoretical complexity for the hydrogenic abstraction reaction. *Mol. Phys. 109*, 2353.

[154] Molina-Espíritu, M.; Esquivel, R. O.; Angulo, J. C.; Antolín, J.; Dehesa, J. S. (2012). Information-theoretical complexity for the hydrogenic identity s_N2 exchange reaction. *J. Math. Chem. 50*, 1882.

[155] Esquivel, R. O.; Molina-Espíritu, M.; Dehesa, J. S.; Angulo, J. C.; Antolín, J. (2012). Concurrent phenomena at the transition region of selected elementary chemical reactions: An informationtheoretical complexity analysis. *Int. J. Quant. Chem. 112*, 3578.

[156] Kullback, S.; Leibler, R. A. (1951). On information and sufficiency. *Ann. Math. Stat. 22*, 79.

[157] Kullback, S. (1959). *Information Theory and Statistics*. New York: John Wiley & Sons.

[158] Sagar, R. P.; Guevara, N. L. (2008). Relative entropy and atomic structure. *J. Mol. Struct. (Theochem) 857*, 72.

[159] Nagy, Á.; Romera, E. (2009). Relative Rényi entropy for atoms. *Int. J. Quant. Chem. 109*, 2490.

[160] Villani, C. (2000). *Topics in Optimal Transportation, in Graduate Studies in Mathematics*, Volume 58. New York: American Mathematical Society.

[161] Yamano, T. (2013). Relative Fisher information for Morse potential and isotropic quantum oscillators. *J. Math. Phys. 54*, 113301.

[162] Sánchez-Moreno, P.; Zarzo, A.; Dehesa, J. S. (2012). Jensen divergence based on Fisher's information. *J. Phys. A 45*, 125305.

[163] Toscani, G. (2017). Poincaré-type inequalities for stable densities. *Ric. Mat. 66*, 15.

[164] Yamano, T. (2013). de bruijn-type identity for systems with flux. *Eur. Phys. J. B 86*, 363.

[165] Frieden, B. R.; Plastino, A.; Plastino, A. R.; Soffer, B. H. (1999). Fisher-based thermodynamics: Its Legendre transform and concavity properties. *Phys. Rev. E 60*, 48.

[166] Antolín, J.; Angulo, J. C.; López-Rosa, S. (2009). Fisher and Jensen-Shannon divergences: Quantitative comparisons among distributions. Application to position and momentum atomic densities. *J. Chem. Phys. 130*, 074110.

[167] Levämäki, H.; Nagy, Á; Vilja, I.; Kokko, K.; Vitos, L. (2017). Kullback-Leibler and relative Fisher information as descriptors of locality. *Int. J. Quant. Chem. 118*, e25557.

[168] Flego, S. P.; Plastino, A.; Plastino, A. R. (2011). Non-linear relativistic diffusions. *Physica A 390*, 2776.

[169] Frieden, B. R.; Soffer, B. H. (2010). Weighted Fisher informations, their derivation and use. *Phys. Lett. A 374*, 3895.

[170] Flego, S. P.; Plastino, A.; Plastino, A. R. (2011). Fisher information, the Hellmann-Feynman theorem, and the Jaynes reciprocity relations. *Ann. Phys. 326*, 2533.

[171] Venkatesan, R. C.; Plastino, A. (2014). Legendre transform structure and extremal properties of the relative Fisher information. *Phys. Lett. A 378*, 1341.

[172] Venkatesan, R. C.; Plastino, A. (2015). Hellmann-Feynman connection for the relative Fisher information. *Ann. Phys. 359*, 300.

[173] Mukherjee, N.; Roy, A. K. (2018). Relative Fisher information in some central potentials. *Ann. Phys. (N.Y.) 398*, 190.

[174] Yamano, T. (2018). Relative Fisher information of hydrogen-like atoms. *Chem. Phys. Lett. 691*, 196.

[175] Laguna, H. G.; Sagar, R. P. (2014). Quantum uncertainties of the confined harmonic oscillator in position, momentum and phasespace. *Ann. Phys. (Berlin) 526*, 555.

[176] Griffiths, D. J. (2004). *Introduction to Quantum Mechanics*. New Jersey: Pearson Prentice Hall.

[177] Mukherjee, N.; Roy, A.; Roy, A. K. (2015). Information entropy as a measure of tunneling and quantum confinement in a symmetric double-well potential. *Ann. Phys. (Berlin) 527*, 825.

[178] Mukherjee, N.; Roy, A. K. (2016). Quantum confinement in an asymmetric double-well potential through energy analysis and information-entropic measure. *Ann. Phys. (Berlin) 528*, 412.

[179] Cohen-Tannoudji, C.; Diu, B.; Laloe, F.; (1978). *Quantum Mechanics*. New York: Wiley.

[180] Gueorguiev, V. G.; Rau, A. R. P.; Draayer, J. P. (2006). Confined one-dimensional harmonic oscillator as a two-mode system. *American Journal of Physics 74*, 394.

[181] Mukherjee, N.; Roy, A. K. (2018). Information−entropic measures in confined isotropic harmonic oscillator. *Adv. Theory Simul. 1*, 1800090.

[182] Yáñez, R. J.; Van Assche, W.; Dehesa, J. S. (1994). Position and momentum information entropies of the D-dimensional harmonic oscillator and hydrogen atom. *Phys. Rev. A 50*, 3065.

[183] Mukherjee, N.; Roy, A. K. (2019). Quantum mechanical virial-like theorem for confined quantum systems. *Phys. Rev. A 99*, 022123.

[184] Roy, A. K. (2014). Confinement in 3D polynomial oscillators through a generalized pseudospectral method. *Mod. Phys. Lett. A 29*, 1450104.

In: Harmonic Oscillators
Editor: Yilun Shang
ISBN: 978-1-53615-810-6

Chapter 2

CLASSICAL-QUANTUM CORRESPONDENCE REVISITED: A CASE STUDY IN THE CONTEXT OF LINEAR SIMPLE HARMONIC OSCILLATOR

***Nabakumar Bera*[1,*]*PhD, Subhasree Ghosh*[2], *PhD and Paulami Ghosh*[3]**
[1]Department of Chemistry, Jhargarm Raj College, West Bengal, India
[2]Department of Chemistry, Diamond Harbour Women's University, West Bengal, India
[3]School of Mathematical and Computational Sciences, Indian Association for the Cultivation of Science, Kolkata, India

1. INTRODUCTION

Physics of macroscopic system that is the large scale phenomena like collisions of two marbels, scattering of visible light radiation by a material particle etc. are described by classical physics. For a system of one or two particles we use Newton's equations of motion. Two other alternatives are Hamiltonian equations of motion and Lanrange's equations of motion[1, 2]. If there is a

*Corresponding Author's E-mail: nabakumarbera@gmail.com.

huge number of particles then we have to use the concept of probability and statistics. In the classical domain, we have to use Maxwell-Boltzmann (MB) statistics[3, 4]. Nature of light or wave as a whole can be best understood by Maxwell's electromagnetic theory. During the last three decades of nineteenth century and the first three decades of twentieth century more and more evidences came to light that the classical framework could not account for several phenomena which seemed to be in violent contradiction with explanations. This includes the phenomena of blackbody radiation[5–8], photoelectric effect[9], temperature dependence of heat capacity of solid, atomic line spectra, Stern-Gerlach experiment[13], Compton effect[14], Davisson-Germer and Thomson's electron diffrction experiment[15–17], Raman effect [10, 11, 18, 19] etc.

Now the question is why classical physics fail to explain these observations?

In classical physics it is taken as granted that energy exchange takes place continuously. But experiments show that there are cases where energy exchange takes place in a discrete or quantized way. When we analyse the experimental phenomena where Classical Mechanics fail, we find that they can be explained only when we consider that energy exchange actually takes place in a discrete way. Therefore we have to develop a systematic approach to find out quantized energy expression of a series of quantum mechanical systems. Bohr-Wilson-Sommerfeld quantization rule is one of such apporaches to find out the quantized energy expression for a class of systems executing a closed periodic orbit. This technique is applicable only for a few specific systems. For arbitrary cases we have to use Schrödinger equation for spin-free nonrelativistic case and Dirac equation for relativistic case. Schrödinger's comment is relavent in this context. *"The great revelation of the quantum theory was the features of discreteness were discovered in the Book of Nature, in a context in which anything oher than continuity seemed to be absurd according to the views held until then."*

Let us go one step back and ask the question: why energy levels are discrete or quantized?

In classical physics it is again taken as granted that position and momentum of a particle can be determined simultaneously with arbitrary precision. But there is an intrinsic principle in nature which says that position and momentum of a particle cannot be determined simultaneously with arbitrary precision. This is the famous Heisenberg's Uncertainty Principle, named after Werner Heisenberg [30]. It is this principle which plays the key role for the peculiar behavior for a particle in the sub-atomic level. This principle resists classical mechanics to be applicable in the microscopic world. When we contruct Schrödinger equation for a bound particle and solve them, then energy that comes out is automatically quantized or discrete.

If we want to investigate the origin of uncertainty principle, we have to consider another important concept in quantum mechanics. This is de Bröglie's wave particle duality [31]. Light shows its wave character in some experiments like interference, diffraction, polarization etc. But in some other experiments light behaves as a particle as in case of photo-electric effect, Compton effect etc. This implies that light has a dual character: particle and wave.

de Bröglie suggested that if light say X-ray, which was originally regarded as wave, has a particle character, why electron, which was regarded as a fundamental particle, will not have a wave character? In 1924 he postulated that for every moving particle there is an associated wave, called the matter wave, having wave length $\lambda = \dfrac{h}{p} = \dfrac{h}{mv}$.

To explain these new case of phenomena a new mathematical tool is essential. This is the Quantum Mechanics. The central equation in quantum mechanics is the Schrödinger equation [26]. This is a linear second order differential equation. An alternative recipee is the Heisenberg's equation of motion which is a matrix equation. These equations are valid in the nonrelativistic case only when particle velocity is much less in comparison to the speed of light. In the relativistic case we have to use relativistic Schrödinger equation, called Dirac equation [28, 29]. Together with these fundamental equations there are several theorems and operator algebra that are useful to treat a quantum mechanical system systematically. Quantum Mechanics can be thought of roughly as the study of physics on very small length scales, although there are also certain macroscopic systems it directly applies to.

Classical mechanics gives the correct results in most of the cases mainly for macroscopic or bulk systems but it fails in microscopic or sub-atomic domain. On the other hand quantum mechanics is complete and it can be used, at least in principle, to microscopic systems as well as macroscopic systems. This implies that behavior of classical objects can be explained by both classical mechanics and the quantum mechanics. Now the question is: how can one decide, without solving the equation, whether a particular problem should be solved by quantum mechanics or classical mechanics is sufficient to explain all the essential physics? One effective approach is as follows:

If the de Bröglie wave length $\lambda \gg d$; where d is the system size, then classical mechanics is sufficient to explain all the essential physics associated with the system. But if $\lambda \lesssim d$; then quantum nature is important and we should solve the system quantum mechanically to explain all the essential physics associated with the system.

Although quantum mechanics is complete but there are only a few simple systems that can be solved exactly. In most of the cases the equations are too complex to be solved exactly. Dirac wrote, *"the underlying physical laws necessary for the mathematical theory of a large part of physics and the whole chemistry are thus completely known, and the difficulty is only that the exact application of these laws leads to equation much too complicated to be soluble. It therefore becomes desirable that approximate practical methods of applying quantum mechanics should be developed, which can lead to an explanation of the main features of complex atomic systems without too much computation."* In such cases we may use various approximate techniques like perturbation theory, variation method etc. After the advent of very powerful computers it is now possible to handle a lot of really challenging problems upto a very high accuracy and this matches with the experimental results.

When we look at the physics of microscopic world we shall certainly apply the principles of quantum mechanics. Classical mechanics is usually used in the macroscopic world. Now the question naturally arises are they totally different or there is a correspondence between classical and quantum mechanical results? We will find that there is certainly a correspondence. In this chapter we will review the classical-quantum correspondence principle from different approaches. For illustration, we will consider different systems where this principle works fine. We shall mainly focus on the linear simple harmonic oscillator (SHO). After revisiting the classical and quantum linear SHO and their different solution techniques we will concentrate and highlight those parts more extensively where correspondence principle can be applicable. We will mainly focus on the zero point energy, probability distribution, quantum mechanical tunneling, dynamics of position, momentum, their higher moments and fluctuations, virial theorem, Hellmann-Feynmann theorem etc. We will try to represent them pedagogically such that even an undegraduates or graduate student in both physics and chemistry community find its relevance in connection to their syllabus.

2. Classical-Quantum Correspondence

Classical mechanics usually gives correct results in the macroscopic or bulk world but it fails in the microscopic or subatomic world. On the other hand quantum mechanics, at least in principle, is general and complete. If the technical complicacy can be resolved then this theory can be used everywhere. Now a question naturally arises what happens if we are able to solve a problem both classically and quantum-mechanically? Are the results are completely different or they have a connection? If we critically examine the results obtained classically and quantum mechanically we find that there is a close connection between these two results. This is what is known as classical quantum correspondence. This term implies that a new theory (quantum mechanics) should reproduce under some condition the results of older well-established theories (classical mechanics) in those cases where older theories work.

This correspondence can be visualized from different ends. The famous one is the Bohr's correspondence principle. According to this principle quantum mechanically calculated results match with the classically expected result when quantum number is very high. We will see how this principle works in reality by taking some simple exactly solvable problems. Another approach is to consider the action integral $S = \int p_i dq_i$, where q_i and p_i are generalized position and momentum coordinates. If the Planck's constant $h \ll S$, then quantum mechanical results reduce to the classical results.

When we consider a quantum mechanical system then there are two length scales which are very important. First one is the de-Bröglie wavelength (λ) and other one is the system size (d). It is observed that if $\lambda \gtrsim d$ then quantum mechanics is important and it may give results different from the results obtained classically. But if $\lambda \ll d$ then the quantum mechanical results will match with the classical results. In this condition we can safely treat the system classically. Under these approximation Schrödinger equation reduces to classical Hamilton-Jacobi equation.

As far as dynamics is concerned we cannot use Newton's equation of motion and it is not possible to get a well defined phase space trajectory. This is due to the dominance of Heisenberg's uncertainty principle in the microscopic world. The usual technique is to use Ehrenfest equation of motion [38]. This is a quantum analog of Classical Newton's equation of motion. The only difference is the instantaneous position and momentum have to be replaced by their average values. This is another important aspect where classical quantum correspondence can be visualized. There are many such aspects where classical-quantum correspondence are quite apparent. Some of them are being described in brief in this article.

3. Planck's Constant $h \to 0$ and the Classical-Quantum Correspondence

To illustrate the fact that as $h \to 0$ then quantum mechanical results reduce to the classical result let us consider the phenomenon of black-body energy distribution. From Planck's formula [5–8], we get the energy density within the frequency range ν and $\nu + d\nu$ and it is given by

$$\rho(\nu)d\nu = \frac{8\pi h\nu^3}{c^3} \frac{d\nu}{\left(e^{\frac{h\nu}{k_{\mathrm{B}}T}} - 1\right)} \tag{1}$$

where k_{B} is the Boltzmann constant, c is the speed of light and T is the temperature in Kelvin.

At a given temperature if we assume that $h \to 0$, then we can write $\frac{h\nu}{k_{\mathrm{B}}T} \to 0$. Then,

$$e^{\frac{h\nu}{k_{\mathrm{B}}T}} \simeq 1 + \frac{h\nu}{k_{\mathrm{B}}T}$$

$$\text{and} \quad \rho(\nu)d\nu = \frac{8\pi h\nu^3}{c^3} \frac{d\nu}{\left(\frac{h\nu}{k_{\mathrm{B}}T}\right)} = \frac{8\pi k_{\mathrm{B}}T\nu^2}{c^3} d\nu \tag{2}$$

This is the Rayleigh-Jeans formula for black-body energy distribution as obatined classically using the equipartition principle. This is due to the fact that as $h \to 0$ discrete energy levels becomes continuous. For classical oscillator, energy levels are continuous. For quantum oscillator, energy levels are discrete and the spacing between two successive energy levels is $h\nu$. If $h \to 0$, then this spacing also tends to zero. This is the result as expected classically.

Another interesting point where this correspondence can be apparent. Single particle Micro-cannonical partition function for a linear SHO is given by

$$Z = \frac{e^{-\frac{h\nu}{2k_{\mathrm{B}}}T}}{1 - e^{-\frac{h\nu}{k_{\mathrm{B}}T}}}$$

and the average energy is given by

$$\langle E\rangle = k_{\mathrm{B}}T^2\left(\frac{\partial \ln Z}{\partial T}\right) = k_{\mathrm{B}}T^2\frac{\partial}{\partial T}\left[\ln\left\{\frac{e^{-\frac{h\nu}{2k_{\mathrm{B}}T}}}{1-e^{-\frac{h\nu}{k_{\mathrm{B}}T}}}\right\}\right]$$

$$\begin{aligned}
\text{or,}\quad \langle E\rangle &= k_{\mathrm{B}}T^2\frac{\partial}{\partial T}\left[-\frac{h\nu}{2k_{\mathrm{B}}T}-\ln\left(1-e^{-\frac{h\nu}{k_{\mathrm{B}}T}}\right)\right]\\
&= k_{\mathrm{B}}T^2\left[\frac{h\nu}{2k_{\mathrm{B}}T^2}-\frac{(-1)\,e^{-\frac{h\nu}{k_{\mathrm{B}}T}}\left(+\frac{h\nu}{k_{\mathrm{B}}T^2}\right)}{1-e^{-\frac{h\nu}{k_{\mathrm{B}}T}}}\right]\\
&= \frac{1}{2}h\nu+\frac{h\nu}{e^{\frac{h\nu}{k_{\mathrm{B}}T}}-1}
\end{aligned}$$

Now when $h \rightarrow 0$ then $\frac{h\nu}{k_{\mathrm{B}}T} \ll 1$ and $e^{\frac{h\nu}{k_{\mathrm{B}}T}} \simeq 1+\frac{h\nu}{k_{\mathrm{B}}T}$. Average energy now becomes

$$\langle E\rangle = k_{\mathrm{B}}T$$

This is the value obtained classically using the equipartition principle.

4. Bohr's Correspondence Principle

Bohr's correspondence principle demands that when the system is very large then classical mechanics and quantum mechanics give the same result. The conditions under which quantum mechanical results match with the classical results are referred to as correspondence limit. In 1920, Bohr provided a rough prescription for the correspondence limit. It states that when the quantum number describing the system is very high then quantum mechanical results match with the classically expected results.

In the Schrödinger's representation, quantum mechanics only reproduces classical mechanics in the probabilistic or statistical interpretation and the statistical interpretation only gives that probabilities of different classical outcomes. Bohr has argued that quantum physics does not reduce to classical physics in the same way as relativistic mechanics reduces to nonrelativistic mechanics when velocity of the particle is much less in comparison to speed of light or wave optics reduces to geometrical optics when wavelength $\lambda \to 0$. He argued that classical physics exists independently of quantum mechanics and cannot be derived from it.

Although the correspondence principle was formulated by Neils Bohr in 1920 [37] but he had made use of it in 1913 when Bohr theory for hydrogen atom or hydrogen-like one electron ions was developed [20–23]. Let us illustrate Bohr's atomic model in brief and see how corresponding principle evolves in this. We know that one electron problems such as H, He^+ etc. are central force problem for which angular momentum is constant. Bohr assumed that atomic orbits are circular. If the electron moves with a velocity v along circular orbit of radius r then angular momentum

$$L = mvr \quad = \quad k \ \ (\text{say}) \tag{3}$$

where k is a constant. Here we have assumed that the nucleus being much heavier than electron is kept stationary. For the mechanical force balance, coulombic attraction force will be cancelled by the centrifugal force i.e.,

$$\frac{ze^2}{r^2} \quad = \quad \frac{mv^2}{r} \tag{4}$$

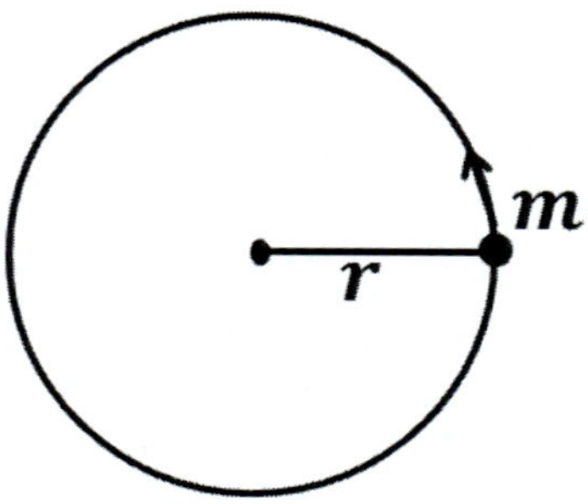

Figure 1. A particle of mass m is rotating on a circular orbit of radius r.

From equations (3) and (4),

$$r = \frac{k^2}{mze^2} \tag{5}$$

Now total energy of the electron

$$\begin{aligned} E = T + V &= \frac{1}{2}mv^2 - \frac{ze^2}{r} \\ &= \frac{ze^2}{2r} - \frac{ze^2}{r} \quad \text{[using equation (4)]} \\ &= -\frac{ze^2}{2r} \\ \text{or,} \quad E &= -\frac{mz^2e^4}{2k^2} \quad \text{[Using equation (5)]} \end{aligned} \tag{6}$$

From experiment it is observed that atomic spectra is a line spectra which indicates that energy levels are discrete or quantized. From Balmer-Rydberg equation and Ritz combination principle [12, 42], it is clear that emission must be in quanta of light. Moreover it must be inversely proportional to $\frac{1}{n^2}$ where n is an integer. As frequency $\nu \propto \Delta E$, it clearly indicates that k must be directly proportional to n. If we write $k = fn$ the from equation (6),

$$E_n = -\frac{mz^2e^4}{2f^2n^2} \tag{7}$$

Here f is a proportionality constant. Frequency of transition between two successive energy levels is given by

$$\begin{aligned} \nu_Q = \frac{E_{n+1} - E_n}{h} &= \frac{mz^2e^4}{2f^2h}\left[\frac{1}{n^2} - \frac{1}{(n+1)^2}\right] \\ &= \frac{mz^2e^4}{2f^2h}\left[\frac{(n+1)^2 - n^2}{n^2(n+1)^2}\right] \\ \text{or,} \quad \nu_Q &= \frac{mz^2e^4}{2f^2h}\frac{(2n+1)}{n^2(n+1)^2} \end{aligned} \tag{8}$$

If we consider classical frequency then it should be

$$\nu_{\text{cl}} = \frac{v}{2\pi r} = \frac{k}{2\pi m r^2} \quad \text{[Using equation (3)]}$$
$$= \frac{m z^2 e^4}{2\pi k^3} \quad \text{[Using equation (5)]}$$
$$\text{or,} \quad \nu_{\text{cl}} = \frac{m z^2 e^4}{2\pi f^3 n^3} \tag{9}$$

From equations (8) and (9) we see that the classical and quantum frequencies are different. Now let us find what happens when $n \to \infty$. From equation (8), it is clear that in the limit $n \to \infty$,

$$\nu_{\text{Q}} = \frac{m z^2 e^4}{f^2 h n^3} \tag{10}$$

If we make a bold assumption that when $n \to \infty$, quantum mechanical frequency (ν_{Q}) be equal to the classical frequency (ν_{cl}) then from equations (9) and (10) it is clear that

$$f = \frac{h}{2\pi} \tag{11}$$

This means angular momentum

$$L = n \frac{h}{2\pi} \quad \text{with } n = 1, 2, 3, \cdots \tag{12}$$

This is the Bohr's quantization rule.

From this analysis it is clear that everything are in order if we assume that in the high quantum number limit quantum mechanical frequency reduces to the classical results. This is what is known as Bohr's correspondence principle.

To illustrate this principle further let us consider a free particle of mass m confined in a one dimensional box of length L. Energy eigen value for this system can be written as

$$E_n = \frac{n^2 h^2}{8 m L^2} \quad \text{with } n = 1, 2, 3, \cdots \tag{13}$$

The quantum mechanical frequency is given by

$$\nu_{\text{Q}} = \frac{\Delta E}{h} = \frac{E_{n+1} - E_n}{h}$$
$$\text{or,} \quad \nu_{\text{Q}} = (2n+1) \frac{h}{8 m L^2} \tag{14}$$

As the particle is free inside the box, its potential energy is zero and total energy is solely the kinetic energy. Therefore, velocity of the particle

$$v = \sqrt{\frac{2E_n}{m}} = \sqrt{\frac{2n^2h^2}{8m^2L^2}}$$
$$\text{or,} \quad v = \frac{nh}{2mL} \tag{15}$$

The classical frequency is given as

$$\nu_{\text{cl}} = \frac{v}{2L} = \frac{nh}{4mL^2} \tag{16}$$

From equations (14) and (16), it is clear that when $n \to \infty$ $\nu_{\text{Q}} = \nu_{\text{cl}}$. This is in accordance with Bohr's correspondence principle.

Let us consider the rigid rotor problem which is another exactly solvable problem. This is equivalent to the problem of a free particle moving on the surface of a sphere. Its energy eigen value is given as

$$E_J = \frac{J(J+1)h^2}{8\pi^2 I} \tag{17}$$

when $J = 0, 1, 2, 3, \cdots$ are the rotational quantum number, I is the moment of inertia. The quantum mechanical frequency is given as

$$\nu_{\text{Q}} = \frac{\Delta E}{h} = [(J+1)(J+2) - J(J+1)]\frac{h}{8\pi^2 I}$$
$$\text{or,} \quad \nu_{\text{Q}} = (2J+1)\frac{h}{8\pi^2 I} \tag{18}$$

For this system, total energy is the rotational kinetic energy. If ω be the angular velocity then

$$\frac{1}{2}I\omega^2 = E_J = \frac{J(J+1)h^2}{8\pi^2 I}$$
$$\text{or,} \quad \omega = \sqrt{\frac{J(J+1)h^2}{4\pi^2 I^2}}$$
$$= \sqrt{J(J+1)}\frac{h}{2\pi I} \tag{19}$$

Hence classical frequency is given as

$$\nu_{\text{cl}} = \frac{\omega}{2\pi} = \frac{\sqrt{J(J+1)}h}{4\pi^2 I} \tag{20}$$

when quantum number is very high

$$\nu_{\rm Q} \simeq \frac{Jh}{4\pi^2 I} \quad = \quad \nu_{\rm cl}$$

This is again in accordance with the Bohr's correspondance principle.

In the above examples we have mainly concentrated on a particular aspect of correspondence principle where quantum mechanical frequency reduces to classical frequency in the high quantum number limit. This principle is manifested in other properties also. Moreover alternative visualizations of correspondance principle are illustrated considering another exactly solvable problem- the linear simple harmonic oscillator.

5. Linear Simple Harmonic Oscillator (LSHO)

The linear Simple Harmonic Oscillator is one of the most important model problems in classical mechanics as well as quantum mechanics for the following reasons:

1. It is exactly solvable both classically and quantum mechanically and a simple solution exists for it.

2. A wide variety of physical situations can be reduced to it either exactly or approximately. It can be used to approximte an arbitrary continuous binding potential in the vicinity of a stable equilibrium position. For systems in which there exist small vibrations about a point of stable equilibrium, oscillations can be approximated to be simple harmonic and system can be treated as a linear harmonic oscillator.

3. The linear SHO problem is the foundation for the understanding of complex modes of vibration in large molecules, the motion of atoms in a solid lattice, (the theory of heat capacity), vibrational motion of nuclei in molecules, quantum theory of light etc.

5.1. Classical Linear SHO Revisited

In classical mechanics a linear SHO is a system of mass m which moves in one dimension in such a way that it is attracted to a fixed center by a force proportional to the diplacement from the centre. This is the Hooke's law. Choosing origin as the centre of force, the linear restoring force is given by

$$F = -kx \tag{21}$$

where k is a positive constant called the force constant, which is a measure of stiffness or rigidity of the spring and represents the restoring force per unit displacement. Then the potential energy

$$V(x) = \frac{1}{2}kx^2 = \frac{1}{2}m\omega^2x^2 = 2\pi^2m\nu^2x^2$$

with $\omega = \sqrt{\frac{k}{m}}$ called the angular frequency of oscillator. Then the equation of

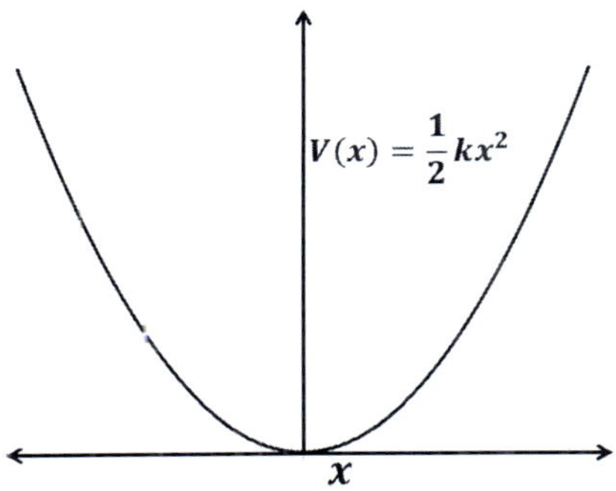

Figure 2. Potential energy for linear Simple Harmonic Oscillator.

motion is

$$\frac{d^2x}{dt^2} + \omega^2x = 0 \tag{22}$$

where x is a function of time. The solution of this equation is oscillatory i.e. harmonic and may be expressed as sinusoidal or cosinusoidal. Assuming the initial phase as zero the solution is

$$x(t) = a\sin(\omega t) \tag{23}$$

with time period $\tau = \frac{2\pi}{\omega} = \frac{1}{\nu}$. a is the maximum amplitude. Velocity $(v(t))$ and acceleration $(f(t))$ are given by

$$v(t) = a\omega\cos(\omega t) = \omega\sqrt{a^2 - x^2} \tag{24}$$

$$f(t) = -\omega^2x \tag{25}$$

Kinetic energy and potential energy of the particle are given by

$$T = \frac{1}{2}m\omega^2 a^2 \cos^2(\omega t) = \frac{1}{2}m\omega^2\left(a^2 - x^2\right) \quad (26)$$

$$V = \frac{1}{2}m\omega^2 a^2 \sin^2(\omega t) = \frac{1}{2}m\omega^2 x^2 \quad (27)$$

and the total energy

$$E = \frac{1}{2}m\omega^2 a^2 \quad (28)$$

This is constant at any time and hence it is a conservative system. Average kinetic and potential energy are given by

$$\langle T\rangle = \langle V\rangle = \frac{E}{2} \quad (29)$$

This is in accordance with the virial theorem.

For classical oscillator energy levels are continuous and zero point vibrational energy is zero. From equation (24), it is clear that the velocity of the particle is maximum at the equilibrium point $(x = 0)$ which is equal to $v_{\max} = a\omega$ and it is zero when $x = \pm a$. These points are called the classical turning points. At the classical turning points, kinetic energy is zero and total energy is fully the potential energy. On the other hand at the equilibrium position, potential energy is zero and total energy is solely the kinetic energy.

As velocity is maximum at the equilibrium position least will be the time it will stay at the equilibrium position. Consequently, probability density will be minimum at the equilibrium position and it will increase as we approach the classical turning points. Another important point that should be noted that there is a zero probability of finding the particle outside the classical turning points. Otherwise velocity of the particle will be imaginary and kinetic energy will be higher which are absurd.

5.2. Solution of Linear SHO Using Old Quantum Theory

Journey of quantum theory has been started on 14th December, 1900 when Max Planck explained the black-body energy distribution curves from theoretical point of view. In this theory he made the revolutionary assumption that energy exchange by atomic oscillators in a solid may take place in a discrete or quantized way. Journey towards quantum theory attain its maturity around 1925 when Schrödinger, Heisenberg, Dirac etc developed a systematic mathematical tool to handle the particles in microscopic or subatomic level. This is what is known as modern quantum mechanics. In between these time frames a set of realistic correction to classical mechanics were made which is collectively known as old quantum theory.

This main tool of old quantum theory is the Wilson-Sommerfeld quantization rule [24, 25] (WSQR). According to this rule for a closed and periodic orbit

$$\oint p_i dq_i = n_i h \text{ with } n_i = 1, 2, 3, \cdots \tag{30}$$

where q_i's are the generalized position coordinate and p_i's are the generalized momentum coordinate.

WSQR is reduced to Bohr's quantization rule when angular momentum is constant. When electron rotates along a circular orbit of fixed radius then we can write

$$\oint L d\phi = nh \tag{31}$$

where L is the angular momentum and ϕ is the angular position. For a central force problem, L is constant. Hence,

$$\begin{aligned} L \oint d\phi &= nh \\ \text{or, } L \times 2\pi &= nh \\ \text{or, } L &= n\frac{h}{2\pi} \quad \text{with } n = 1, 2, 3 \cdots \end{aligned} \tag{32}$$

This is the Bohr's quantization rule.

Another interesting results can be obtained if linear momentum p is constant. Then WSQR reduces to

$$p \oint dq = nh \text{ or, } \oint dq = \frac{nh}{p} \tag{33}$$

From de-Bröglie's wave particle duality, wave length associated to the particle $\lambda = \frac{h}{p}$. Hence,

$$\oint dq = n\lambda \text{ with } n = 1, 2, 3, \cdots \tag{34}$$

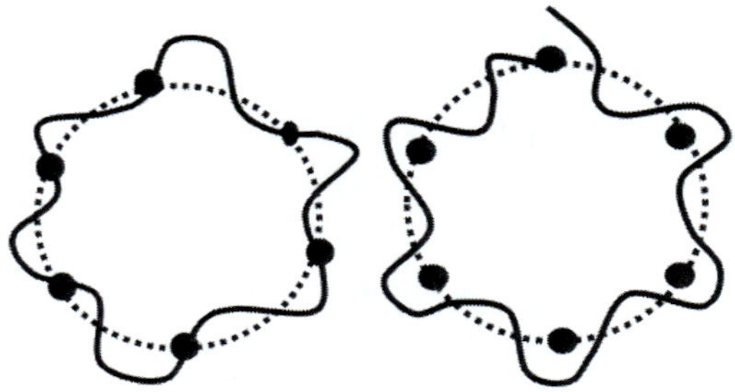

Figure 3. (a) Allowed orbit. (b) Forbidden orbit.

This means within the circumstance of the orbit there must fit an integral number of waves. This is the de-Bröglie's quantization rule. If this condition is satisfied then only we get a stationary orbit. From the following figures the first one is allowed but the second one is forbidden. This rule has been used to solve many problems such as a free particle confined in a box, rigid rotor problem, hydrogen atom problem etc. Here we will use this rule to solve the linear SHO problem. For linear SHO, total energy is given as

$$E = \frac{p_x^2}{2m} + \frac{1}{2}kx^2 \tag{35}$$

$$\text{or,} \quad \frac{x^2}{\left(\sqrt{\frac{2E}{k}}\right)^2} + \frac{p_x^2}{\left(\sqrt{2mE}\right)^2} = 1 \tag{36}$$

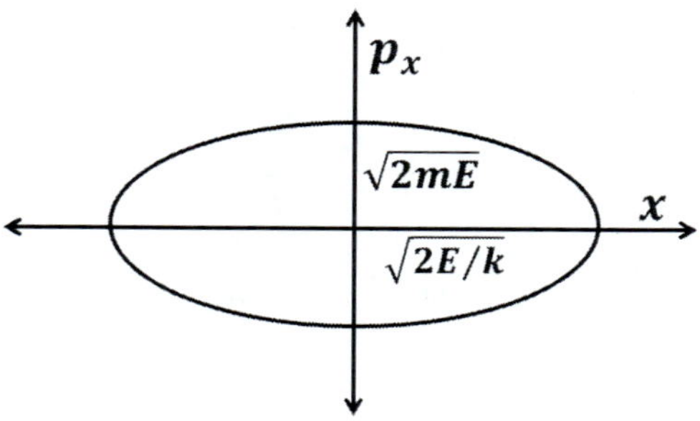

Figure 4. Phase space of a linear SHO.

Equation (36) represents an ellipse with semimajor and semi-minor axes $A = \sqrt{\frac{2E}{k}}$ and $B = \sqrt{2mE}$. A plot where the independent axes are position and momentum is called a phase space. Now the area enclosed,

$$\oint p_x dx = \pi AB \text{ or, } nh = \pi \times \sqrt{\frac{2E}{k}} \times \sqrt{2mE}$$

$$\text{or, } E = \frac{nh}{2\pi}\sqrt{\frac{k}{m}} \text{ or, } E_n = n\hbar\omega \text{ where } \omega = \sqrt{\frac{k}{m}} \text{ and } \hbar = \frac{h}{2\pi}$$

This equation can be alternatively solved as follows. From equation, (35) it is clear that the momentum component can be expressed as

$$p_x = \sqrt{2m\left(E - \frac{1}{2}kx^2\right)} \tag{37}$$

Again total energy $E = \frac{1}{2}ka^2$ where a is the classical turning point. Then according to WSQR,

$$4\int_0^a \sqrt{mk(a^2 - x^2)}\, dx = nh \tag{38}$$

From this equation we can get the value of a^2 and hence total energy E.

Major disadvantage of this WSQR is that it does not incorporate the effect of uncertainty principle. Consequently, the final results may not be absolutely correct in many cases, particularly the non-zero zero point energy which should be zero.

5.3. Quantum Linear SHO

Time independependent Schrödinger equation for linear SHO is

$$\left[-\frac{\hbar^2}{2m}\frac{d^2}{dx^2} + \frac{1}{2}m\omega^2 x^2\right]\psi(x) = E\psi(x) \tag{39}$$

This can be solved either algebraically using ladder operator or by series method. In the series solution equation (39) is first transformed in a standard linear second order differential equation called the Hermite equation

$$\frac{d^2H(y)}{dy^2} - 2y\frac{dH(y)}{dy} + \left(\frac{\lambda}{\alpha} - 1\right) H(y) = 0 \quad (40)$$

where $y = \sqrt{\alpha}\, x$ and $\alpha = \frac{\sqrt{k\mu}}{\hbar} = \frac{m\omega}{\hbar}$. The standard and general method for solving this equation is the so called series method or Frobenius method. After solving this equation we get the energy eigen values and the eigen functions as

$$E_n = \left(n + \frac{1}{2}\right) \hbar\omega \quad (41)$$

$$\text{and} \quad \psi_n(y) = N_n H_n(y) e^{\frac{-y^2}{2}} \quad (42)$$

with $n = 0, 1, 2, 3, \cdots$ are the vibrational quantum number. Here $H_n(y)$ is the Hermite Polynomial and given by

$$H_n(y) = (-1)^n \frac{d^n}{dy^n} \left(e^{-y^2}\right) e^{y^2} \quad (43)$$

and N_n is the normalisation constant,

$$N_n = \frac{1}{\sqrt{2^n n!}} \left(\frac{\alpha}{\pi}\right)^{\frac{1}{4}} \quad (44)$$

Substituting y by $\sqrt{\alpha}x$ in equation (42), we get

$$\psi_n(x) = N_n\, H_n\left(\sqrt{\alpha}x\right)\, e^{\frac{-\alpha x^2}{2}} \quad (45)$$

Using the equations above, we can evaluate the normalised wave function for

$n = 0, 1, 2, 3.....$ and first ten states are given below:

$$\text{For } n = 0, \psi_0 = \left(\frac{\alpha}{\pi}\right)^{\frac{1}{4}} e^{\frac{-\alpha x^2}{2}}$$

$$\text{For } n = 1, \psi_1 = \sqrt{2}\left(\frac{\alpha}{\pi}\right)^{\frac{1}{4}} \sqrt{\alpha} x e^{\frac{-\alpha x^2}{2}}$$

$$\text{For } n = 2, \psi_2 = \frac{1}{2\sqrt{2}}\left(\frac{\alpha}{\pi}\right)^{\frac{1}{4}} \left(4\alpha x^2 - 2\right) e^{\frac{-\alpha x^2}{2}}$$

$$\text{For } n = 3, \psi_3 = \frac{1}{\sqrt{3}}\left(\frac{\alpha}{\pi}\right)^{\frac{1}{4}} \sqrt{\alpha} x \left(2\alpha x^2 - 3\right) e^{\frac{-\alpha x^2}{2}}$$

$$\text{For } n = 4, \psi_4 = \frac{1}{8\sqrt{6}}\left(\frac{\alpha}{\pi}\right)^{\frac{1}{4}} \left(16\alpha^2 x^4 - 48\alpha x^2 + 12\right) e^{\frac{-\alpha x^2}{2}}$$

$$\text{For } n = 5, \psi_5 = \frac{1}{2\sqrt{15}}\left(\frac{\alpha}{\pi}\right)^{\frac{1}{4}} \alpha x^2 \left(4\alpha^2 x^4 - 20\alpha x^2 + 15\right) e^{\frac{-\alpha x^2}{2}}$$

$$\text{For } n = 6, \psi_6 = \frac{1}{96\sqrt{5}}\left(\frac{\alpha}{\pi}\right)^{\frac{1}{4}} \left(64\alpha^3 x^6 - 480\alpha^2 x^4 + 720\alpha x^2 - 120\right) e^{\frac{-\alpha x^2}{2}}$$

$$\text{For } n = 7, \psi_7 = \frac{1}{96\sqrt{70}}\left(\frac{\alpha}{\pi}\right)^{\frac{1}{4}} \sqrt{\alpha} x \left(128\alpha^3 x^6 - 1344\alpha^2 x^4 + 3360\alpha x^2 - 1680\right) e^{\frac{-\alpha x^2}{2}}$$

For $n = 8$,

$$\psi_8 = \frac{1}{384\sqrt{70}}\left(\frac{\alpha}{\pi}\right)^{\frac{1}{4}} \left(256\alpha^4 x^3 - 3584\alpha^3 x^6 + 13440\alpha^2 x^4 - 13440\alpha x^2 + 1680\right) e^{\frac{-\alpha x^2}{2}}$$

For $n = 9$,

$$\psi_9 = \frac{1}{2304\sqrt{35}}\left(\frac{\alpha}{\pi}\right)^{\frac{1}{4}} \sqrt{\alpha} x \left(512\alpha^4 x^8 - 9216\alpha^3 x^6 + 48384\alpha^2 x^4 - 80640\alpha x^2 + 30240\right) e^{\frac{-\alpha x^2}{2}}$$

Let us now plot the wave function as well as probability density and they are given in Figure (5). From the figures it is clear that

1. When $n = 0, 2, 4, \cdots$ (even integer) wave functions are even functions or symmetric about $x = 0$. When $n = 1, 3, 5, \cdots$ (odd integer) then wave function is odd function or anti symmetric about $x = 0$. For linear SHO, wave functions belong to a definite parity as the potential energy $V(x) = \frac{1}{2}kx^2$ is an even function of x.

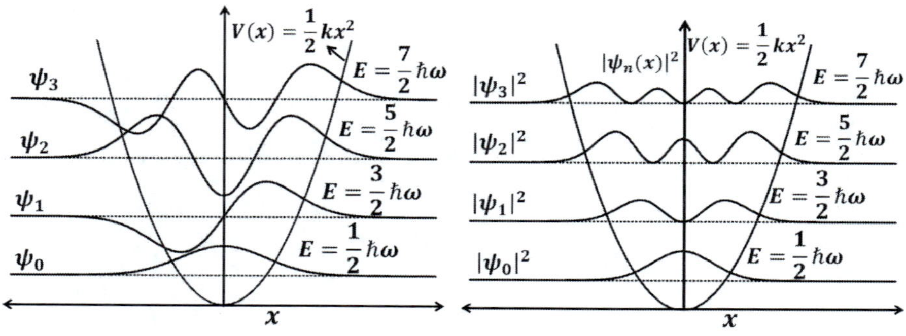

Figure 5. Wavefunctions and probability distribution of linear Simple Harmonic Oscillator.

2. As the system is one dimensional and potential energy is nonsingular and $\psi_n(x) \to 0$ as $x \to \pm\infty$, there is no degeneracy.

3. As far as position and number of nodes are considered the following observations are quite aparent:

   (a) ground state is nodeless.

   (b) nth energy level has $(n-1)$ nodes.

   (c) except the ground state there is a node in the nth state in between two successive nodes of $(n+1)$th state.

From the expression of energy eigen values, it is clear that vibrational energy is discrete or quantized. Energy levels are equispaced and spacing is $\hbar\omega$. Minimum possible energy i.e., the zero point energy is nonzero and it is equal to $\frac{1}{2}\hbar\omega$.

The alternative method of solving equation (39) is the ladder operator method. In this approach we first introduce to operators $\hat{a}_+$ and $\hat{a}_-$ which are defined as

$$\hat{a}_+ = \frac{1}{\sqrt{2\hbar m\omega}}(-i\hat{p}_x + m\omega\hat{x}) \tag{46}$$

$$\text{and} \quad \hat{a}_- = \frac{1}{\sqrt{2\hbar m\omega}}(i\hat{p}_x + m\omega\hat{x}) \tag{47}$$

These operators are linear and obey the commutation relation

$$[\hat{a}_-, \hat{a}_+] = 1 \tag{48}$$

Again $\hat{a}_+$ and $\hat{a}_-$ are hermitian conjugate to each other i.e.,

$$\hat{a}_\pm^\intercal = \hat{a}_\mp \tag{49}$$

For a linear SHO, the important quantum mechanical operators can be expressed in terms of $\hat{a}_+$ and $\hat{a}_-$ as follows.

$$\hat{H} = \hbar\omega\left(\hat{a}_-\hat{a}_+ - \frac{1}{2}\right) = \hbar\omega\left(\hat{a}_+\hat{a}_- + \frac{1}{2}\right) \tag{50}$$

$$\hat{x} = \sqrt{\frac{\hbar}{2m\omega}}\left(\hat{a}_+ + \hat{a}_-\right) \tag{51}$$

$$\hat{p}_x = i\sqrt{\frac{\hbar m\omega}{2}}\left(\hat{a}_+ - \hat{a}_-\right) \tag{52}$$

Other operators can be easily constructed from these fundamental relations. For a linear SHO if $\hat{H}\psi_n = E_n\psi_n$, then,

$$\hat{H}\left(\hat{a}_-\psi_n\right) = \left(E_n + \hbar\omega\right)\left(\hat{a}_+\psi_n\right) \tag{53}$$

$$\text{and}\quad \hat{H}\left(\hat{a}_-\psi_n\right) = \left(E_n - \hbar\omega\right)\left(\hat{a}_-\psi_n\right) \tag{54}$$

This means $\hat{a}_+$ or $\hat{a}_-$ operator acts on a function and forms a new function having energy one unit ($\hbar\omega$) higher or lower. Hence they are called ladder operator. $\hat{a}_+$ is called the raising operator or step-up operator and $\hat{a}_-$ is called the lowering operator or step down operator. When $\hat{a}_-$ operates on the ground state wave function then it must vanish as there is no state below the ground state i.e.,

$$\hat{a}_-\psi_0 = 0 \tag{55}$$

This is the basis of constructing the ground state wave function and ground state energy. They are

$$\psi_0(x) = \left(\frac{\alpha}{\pi}\right)^{\frac{1}{4}} e^{-\dfrac{\alpha x^2}{2}} \tag{56}$$

$$\text{and}\quad E_0 = \frac{1}{2}\hbar\omega \tag{57}$$

Excited states can be easily obtained by operating $\hat{a}_+$ successively on the ground state wave function. The rules are

$$\hat{a}_+\psi_n = \sqrt{n+1}\psi_{n+1} \tag{58}$$
$$\hat{a}_-\psi_n = \sqrt{n}\psi_{n-1} \tag{59}$$

Using this technique we can easily calculate properties associated to a linear SHO for arbitrary states. As for example, we can determine $\langle x\rangle$, $\langle p_x\rangle$, $\langle x^2\rangle$, $\langle p_x^2\rangle$, $\langle T\rangle$, $\langle V\rangle$ etc. They come out as

$$\begin{aligned}
\langle x\rangle = 0,\ \langle p_x\rangle &= 0\\
\langle x^2\rangle = \frac{\hbar}{m\omega}\left(n+\frac{1}{2}\right),\ \langle p_x^2\rangle &= m\hbar\omega\left(n+\frac{1}{2}\right)\\
\langle T\rangle = \frac{1}{2}\left(n+\frac{1}{2}\right)\hbar\omega,\ \langle V\rangle &= \frac{1}{2}\hbar\omega\left(n+\frac{1}{2}\right)
\end{aligned}$$

It is clear that virial equation is automatically obeyed for the linear SHO. Here a point to note that evaluation of $\langle x^2\rangle$, $\langle p_x^2\rangle$ for arbitrary nth state is quite simple using the properties of ladder operators. But if we want to evaluate them using the functional form of the wave function then it will be very tedious particularly for large n.

5.4. Differences of Classical and Quantum Linear SHO

Classical and quantum linear SHO differ in many aspects. Some of them are mentioned below:

1. For classical linear SHO energy levels are continuous but for quantum SHO energy levels are discrete or quantized (see section 1.6).

2. For classical linear SHO, minimum possible energy that is the zero point vibrational energy is zero. This happens when the particle is at rest in its equilibrium position. But for quantum linear SHO zero point energy is nonzero and it is equal to $\frac{1}{2}\hbar\omega$. This is a consequence of of Heisenberg's uncertainty principle (see section 1.7).

3. Probability distribution for quantum SHO is very different from the classical SHO. For classical SHO probability density is maximum near the classical turning points and it is minimum at the equilibrium point. But

for quantum SHO it depends on the quantum number. As for example for the ground state, probability density is maximum at the equilibrium point and it gradually decreases as we approach towards the classical turning points. (This part has been discussed in section (1.9) in greater details.)

4. For classical linear SHO, probability of finding the particle outside the classical turning point is zero. But for quantum oscillator there is a finite probability of finding the particle even outside the classical turning points. This is the phenomenon of quantum mechanical tunneling. (This part has been discussed in section (1.10) in greater details.)

6. Discrete Energy Levels and Classical-Quantum Correspondence

For classical systems energy must be a continuous quantity. But the most important features of quantum mechanical system is that energy is discrete or quantized. As for example, for a hydrogen atom or hydrogen-like one electron system classical energy expression is $E = -\dfrac{ze^2}{2r}$ which is a continuous quantity and the corresponding spectra is expected to be continuous. But atomic spectra is a line spectra. This indicates that energy levels are discrete. From Bohr theory and by solving Schrödinger equation it has been found that for H-like systems energy expression is

$$E_n = -\frac{2\pi^2 mz^2e^4}{n^2h^2} \tag{60}$$

with $n = 1, 2, 3, \cdots$. This expression of energy is in the CGS unit. In the expression for SI unit expression for energy can be obtained by replacing e^2 by $\dfrac{e^2}{4\pi\epsilon_0}$. Spacing between two successive energy levels are

$$\Delta E_n = \frac{2\pi^2 mz^2e^4}{h^2}\frac{(2n+1)}{n^2(n+1)^2} \tag{61}$$

when n is large then

$$\Delta E_n = \frac{4\pi^2 mz^2e^4}{n^3h^3} \tag{62}$$

It is clear that spacing between two successive energy levels decreases with increase in quantum number.

When $n \to \infty$ this spacing becomes zero. This means energy levels become continuous. This is what is expected classically.

But what happens in case of linear SHO? From the expression of energy eigen values, spacing between two successive energy levels $\Delta E = \hbar\omega$ which is constant i.e., independent of quantum number n. This equiapaced energy levels apparently implies that this is against the Bohr's correspondence principle. But this should be visualized differently. If we consider relative energy gap i.e., $\Delta E_n^{\mathrm{rel}} = \dfrac{\Delta E_{(n+1)\leftarrow n}}{E_n}$ then

$$E_n^{\mathrm{rel}} = \frac{\hbar\omega}{\left(n+\frac{1}{2}\right)\hbar\omega} = \frac{1}{\left(n+\frac{1}{2}\right)} \tag{63}$$

When $n \to \infty$, $\Delta E_n^{\mathrm{rel}} \to 0$ which is expected classically. This can be alternatively visualized from the concurrence of classical and quantum frequency. For linear SHO, quantum mechanical frequency $\nu_{\mathrm{Q}} = \dfrac{\Delta E_{(n+1)\leftarrow n}}{h}$ is exactly the same to that of the classical frequency $\nu_{\mathrm{cl}} = \dfrac{\omega}{2\pi} = \dfrac{1}{2\pi}\sqrt{\dfrac{k}{\mu}}$.

For PIAB and rigid rotor problems, the situation is apparently more serious. Here spacing between two successive energy levels increases with increase in quantum number. This certainly does not violate Bohr's corresponding principle. When quantum number tends to infinity then relative energy gap tends to zero and $\nu_{\mathrm{Q}} = \nu_{\mathrm{cl}}$.

For rigid rotor spacing between two successive energy levels

$$\Delta E_{(J+1)\leftarrow J} = (2J+1)\frac{h^2}{8\pi^2 I} \tag{64}$$

This increases with increase in J. Now the relative energy gap

$$\Delta E_J^{\mathrm{rel}} = \frac{(2J+1)\dfrac{h^2}{8\pi^2 I}}{J(J+1)\dfrac{h^2}{8\pi^2 I}} = \frac{2J+1}{J(J+1)} \tag{65}$$

It is clear that when $J \to \infty$, $\Delta E_J^{\text{rel}} \to 0$. For PIAB probelm in 1D, spacing between two successive energy levels is

$$\Delta E_{(n-1)\leftarrow n} = (2n+1)\frac{h^2}{8mL^2}$$

$\Delta E_{(n+1)\leftarrow n}$ increases with increase in n. Now the relative enrgy gap is

$$\Delta E_n^{\text{rel}} = \frac{\Delta E_{(n+1)\leftarrow n}}{E_n} = \frac{(2n+1)\dfrac{h^2}{8mL^2}}{\dfrac{n^2h^2}{8mL^2}} = \frac{(2n+1)}{n^2}$$

It is again clear that as $n \to \infty$, $\Delta E_n^{\text{rel}} \to 0$.

7. Zero Point Energy and Classical-Quantum Correspondence

Zero point energy (ZPE) is the difference of minimum energies of a system when treated quantum mechanically and classically. This is a consequence of Heisenberg's uncertainty principle.

As for example for a free particle of mass m confined in a one dimensional box of length L, its total energy is solely the kinetic energy. If zero point energy be zero then its velocity or momentum (p_x) will be zero. Then uncertainty in momentum

$$\Delta p_x = \sigma_{p_x} = \sqrt{\langle p_x^2\rangle - \langle p_x\rangle^2} = 0 \tag{66}$$

As the particle is confined in a box of finite length its uncertainty in position must be finite. Hence

$$\Delta x \Delta p_x = 0 \tag{67}$$

This violates Heisenberg's uncertainty principle which is absurd. Hence zero point energy cannot be zero. This is one of the reasons why quantum number for PIAB problem starts from $n = 1$ instead of zero. Another reason is that as $n = 0$ then the wave function ψ_n is zero always. This gives a trivial solution having no physical significance. For PIAB problem, zero point energy is

$$E_1 = \frac{h^2}{8mL^2}$$

The relative zero point energy

$$E_1^{\text{rel}} = \frac{E_1}{E_n} = \frac{h^2}{8mL^2 \times \left(\frac{n^2h^2}{8mL^2}\right)} = \frac{1}{n^2} \tag{68}$$

This tends to zero when $n \to \infty$.

For a linear SHO, total energy

$$E = T + V = \frac{p_x^2}{2m} + \frac{1}{2}kx^2 \tag{69}$$

As both terms are square terms, kinetic energy and potential energy both are non-negative. Minimum energy zero implies both $T = 0$, $V = 0$. If $T = 0$, then $\langle p_x^2 \rangle = 0$ and if $\langle V \rangle = 0$, then $\langle x^2 \rangle = 0$. From the symmetry of the problem $\langle x \rangle = 0$ and $\langle p_x \rangle = 0$. Therefore, $\Delta p_x = \sigma_{p_x} = \sqrt{\langle p_x^2 \rangle - \langle p_x \rangle^2} = 0$ and $\Delta x = \sigma_x = \sqrt{\langle x^2 \rangle - \langle x \rangle^2} = 0$.

Therefore the uncertainty product $\Delta x \Delta p_x = 0$ which is absurd. Therefore for a linear SHO, zero point energy must be nonzero. This is clear from the energy expression when $n = 0$ this gives the lowest possible energy i.e., the zero point energy and it is equal to $\frac{1}{2}\hbar\omega$. From the semiclassical treatment (WSQR) $E_n = n\hbar\omega$ with $n = 0, 1, 2, 3, \ldots$. This is not the correct one as it does not obey Heisenberg's uncertainty principle. Here a point to note that when $n \to \infty$ then quantum mechanical results reduces to the semi-classical results.

When we consider the relative zero point energy i.e., $E_0^{\text{rel}} = \frac{E_0}{E_n} = \frac{1}{\left(n + \frac{1}{2}\right)}$. This becomes zero when $n \to \infty$. This is what is expected classically. Therefore, this result is also in agreement with Bohr's correspondence principle.

For a free particle rotating on a ring of fixed radius, for a rigid rotor problem or a free particle on the surface of a sphere, zero point energy is zero. This does not imply that Heisenberg's uncertainty principle is violated. In this case uncertainty in angular position $(\Delta\phi)$ has an intrinsic uncertainty which is arbitrary and tends to infinity as quantum number tends to infinity. Therfore $\Delta L \Delta \phi$ can be $\geq \frac{\hbar}{2}$ even if $\Delta L = 0$.

For a free particle problem, momentum is definite but position is completely arbitrary. This mean $\Delta p_x = 0$ and $\Delta x \to \infty$. In this case also $\Delta x \Delta p_x \geq \frac{\hbar}{2}$ even if $\Delta p_x = 0$.

8. Heisenberg's Uncertainty Principle and Classical-Quantum Correspondence

Apparently strange behavior of particles in the quantum domain is due to position momentum uncertainty principle.

$$\Delta x \Delta p_x \geq \frac{\hbar}{2}$$

Mathematically uncertainty in position and momentum are their respective standard deviations.

$$\begin{aligned}
\Delta x \equiv \sigma_x &= \sqrt{\langle x^2 \rangle - \langle x \rangle^2} \\
\Delta p_x \equiv \sigma_{p_x} &= \sqrt{\langle p_x^2 \rangle - \langle p_x \rangle^2}
\end{aligned}$$

For the linear SHO, symmetry of the problem clearly indicates that

$$\langle x \rangle = \langle p_x \rangle = 0 \tag{70}$$

for any states. For an arbitrary states $\langle x^2 \rangle$ and $\langle p_x^2 \rangle$ can be calculated either using the properties of ladder operators or by using virial theorems. From the functional form of the wave functions it is extremely difficult to get an expression for arbitrary n. From virial theorem, for a linear SHO

$$\langle T \rangle = \langle V \rangle = \frac{\langle E \rangle}{2}$$

Hence,

$$\begin{aligned}
\frac{\langle p_x^2 \rangle}{2m} &= \frac{1}{2}\left(n + \frac{1}{2}\right)\hbar\omega \text{ or, } \langle p_x^2 \rangle = \left(n + \frac{1}{2}\right) m\hbar\omega \\
\text{and } \frac{1}{2} m\omega^2 \langle x^2 \rangle &= \frac{1}{2}\left(n + \frac{1}{2}\right)\hbar\omega \text{ or, } \langle x^2 \rangle = \left(n + \frac{1}{2}\right)\frac{\hbar}{m\omega}
\end{aligned}$$

Therefore, standard deviations of position and momentum are

$$\Delta x = \sigma_x = \sqrt{\left(n+\frac{1}{2}\right)\frac{\hbar}{m\omega}}$$
$$\text{and} \quad \Delta p_x = \sigma_{p_x} = \sqrt{\left(n+\frac{1}{2}\right)m\hbar\omega}$$

Therefore,

$$\Delta x \Delta p_x = \left(n+\frac{1}{2}\right)\hbar \text{ with } n = 0, 1, 2, 3, \cdots \tag{71}$$

Minimum uncertainty product (when $n = 0$) is given as

$$(\Delta x \Delta p_x)_{\text{min}} = \frac{\hbar}{2} \tag{72}$$

This is for the ground state of the linear SHO. We get this minimum uncertainty state as the ground state is Gaussian. The relative uncertainty is given as

$$\frac{(\Delta x \Delta p_x)_{\text{min}}}{(\Delta x \Delta p_x)} = \frac{1}{2\left(n+\frac{1}{2}\right)} \to 0 \text{ when } n \to \infty$$

This is expected classically. This result is agian in agreement with Bohr's correspondence principle. For PIAB problem, similar observation can be found. For this system, uncertainty product is given as

$$\Delta x \Delta p_x = \left(\frac{\pi^2 n^2}{3} - 2\right)^{\frac{1}{2}} \frac{\hbar}{2} \tag{73}$$

When $n = 1$, then this will give the minimum certainty product i.e.,

$$(\Delta x \Delta p_x)_{\text{min}} = \left(\frac{\pi^2}{3} - 2\right)^{\frac{1}{2}} \frac{\hbar}{2} > \frac{\hbar}{2} \tag{74}$$

Now the relative minimum uncertainty

$$\frac{(\Delta x \Delta p_x)_{\text{min}}}{(\Delta x \Delta p_x)} = \left[\frac{\left(\frac{\pi^2}{3} - 2\right)}{\left(\frac{\pi^2 n^2}{3} - 2\right)}\right]^{\frac{1}{2}} \tag{75}$$

This is clear that this relative minimum uncertainty tends to zero as quantum number $n \to \infty$.

9. PROBABILITY DISTRIBUTION AND CLASSICAL-QUANTUM CORRESPONDENCE

An important aspect where quantum oscillator differs from the classical oscillator is the probability distribution. Suppose the classical oscillator spends time dt within the region x and $x + dx$ in a half cycle. Then

$$dt = \frac{dx}{v(x)} \tag{76}$$

where $v(x)$ is the velocity of the particle at point x and it is given by

$$v(x) = \omega\sqrt{a^2 - x^2} \tag{77}$$

Then

$$dt = \frac{dx}{\omega\sqrt{a^2 - x^2}} \tag{78}$$

Now the classical probability of finding the particle within the region x and $x + dx$ is given by

$$P_{\text{cl}}(x)dx = \frac{2dt}{\tau} = \frac{2}{\tau\omega}\frac{dx}{\sqrt{a^2 - x^2}} = \frac{1}{\pi}\frac{dx}{\sqrt{a^2 - x^2}}$$

Or the classical probability density is given by

$$P_{\text{cl}}(x) = \frac{1}{\pi\sqrt{a^2 - x^2}} \tag{79}$$

From this expression it is clear that classical probability density is minimum at $x = 0$, i.e. at the equilibrium position. It increases with increase in $|x|$ and tends to infinity at the classical turning points $x = \pm a$. This is graphically represented in Figure 6.

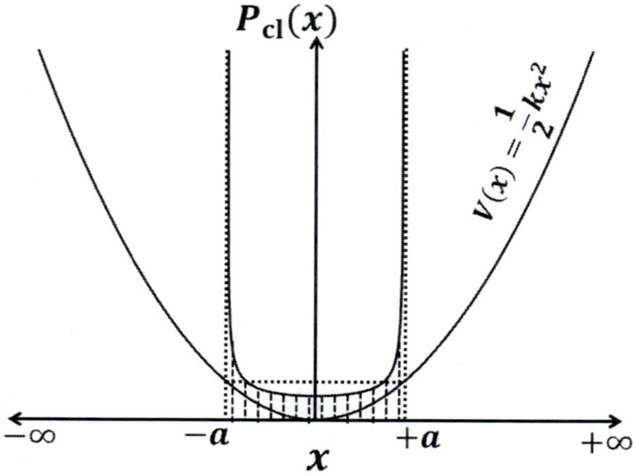

Figure 6. Classical probability distribution of linear SHO.

We know $\frac{1}{2}m\omega^2 a^2 = E = \left(n + \frac{1}{2}\right)\hbar\omega$ or $a^2 = \left(n + \frac{1}{2}\right)\frac{2\hbar}{m\omega}$. Therefore, classical probability density for different n can be written as

$$P_{cl}(x) = \frac{1}{\pi\sqrt{\left(n + \frac{1}{2}\right)\frac{2\hbar}{m\omega} - x^2}} = \frac{1}{\pi\sqrt{\left(n + \frac{1}{2}\right)\frac{2}{\alpha} - x^2}} \quad (80)$$

For the quantum linear SHO probability density at a particular quantum state can be easilily calculated from the expression of wavefunctions as follows

$$P_{\text{quan}}(x) = \psi_n(x)^*\psi_n(x) \quad (81)$$

Quantum probability density for first few states are given below.

$$(P_{\text{quan}})_0(x) = \left(\frac{\alpha}{\pi}\right)^{\frac{1}{2}} e^{-\alpha x^2} \quad (82)$$

$$(P_{\text{quan}})_1(x) = 2\left(\frac{\alpha}{\pi}\right)^{\frac{1}{2}} \alpha x^2 e^{-\alpha x^2} \quad (83)$$

$$(P_{\text{quan}})_2(x) = \frac{1}{8}\left(\frac{\alpha}{\pi}\right)^{\frac{1}{2}} \left[16\alpha^2 x^4 - 16\alpha x^2 + 4\right] e^{-\alpha x^2} \quad (84)$$

$$(P_{\text{quan}})_3(x) = \frac{1}{3}\left(\frac{\alpha}{\pi}\right)^{\frac{1}{2}} \alpha x^2 \left[4\alpha^2 x^4 - 12\alpha x^2 + 9\right] e^{-\alpha x^2} \quad (85)$$

and so on. From the expressions it is clear that for the ground state probability distribution is gaussian in nature. Probability density is maximum at $x = 0$, i.e at the equilibrium position. It decreases and becomes zero when $|x| \to \infty$. For the 1st excited state probability density is zero at the equilibrium position and two maxima appear at $x = \pm\frac{1}{\sqrt{\alpha}}$. For the 2nd excited state maxima appear at $x = 0$ and $x = \pm\frac{1.58}{\sqrt{\alpha}}$ and the minima appear at $x = \frac{0.707}{\sqrt{\alpha}}$. Here an important point to note that the peak height at $x = \pm\frac{1.58}{\sqrt{\alpha}}$ is greater than that at $x = 0$. For the 3rd excited state maxima appear at $x = \pm\frac{0.06}{\sqrt{\alpha}}, \pm\frac{2.04}{\sqrt{\alpha}}$ and the minima appear at $x = 0, \pm\frac{1}{1.225\sqrt{\alpha}}$. In this case also the probability density at $x = \pm\frac{2.04}{\sqrt{\alpha}}$ is greater than that at $x = \pm\frac{0.06}{\sqrt{\alpha}}$. The results have been summarized in table (1) and Figure (7).

Table 1. Comparison of probability maxima for classical and quantum oscillator for $n = 0, 1, 2, 3$

Quantum No. (n)	$(x_{\text{classical}})_{\text{max}}$	$(x_{\text{quantum}})_{\text{max}}$	**% of disagreement**
0	$\pm\frac{1}{\sqrt{\alpha}}$	0	complete disagreement
1	$\pm\frac{1.732}{\sqrt{\alpha}}$	$\pm\frac{1}{\sqrt{\alpha}}$	42%
2	$\pm\frac{2.236}{\sqrt{\alpha}}$	$\pm\frac{1.58}{\sqrt{\alpha}}$	29%
3	$\pm\frac{2.646}{\sqrt{\alpha}}$	$\pm\frac{2.04}{\sqrt{\alpha}}$	23%

From the Figure 7 and the table 1, it is clear that a classical linear SHO is less likely to be found at the equilibrium position $x = 0$ (since the velocity is maximum) and more likely to be found at the ends called classical turning points (since velocity is minimum). From the quantum mechanical probability distribution it is clear that probility density curve has the highest maxima at the ends while the intermediate peaks are low. This indicates that quantum oscillator in excited state is more likely to be found at the ends. This conclusion becomes more and more true for high n cases. Clearly thus for very high quantum numbers quantum results merge with classical results as per Bohr's correspondence principle.

The classical and quantum probability distributions show some striking dissimilarities. While classical curve is continuous or smooth, the quantum curve consists of n nodes and $n+1$ peaks for the nth state. However as n increases the nodes become more and more closely spaced. The quantum curve for high n shows rapid fluctuations (since kinetic energy is high). However the mean curve agrees with the classical one. Similar situation arises for PIAB problem. As the particle is free inside the box of length L, the classical probability distribution will be homogeneous (constant) and independent of position. Classical probability is given by

$$P_{\text{cl}} = \frac{1}{L} \tag{86}$$

whereas quantum probability distribution is given by

$$P_{\text{quan}}(x) = |\psi(x)|^2 = \frac{2}{L}\sin^2\left(\frac{n\pi}{L}x\right) \quad \text{with } n = 1, 2, 3\cdots \tag{87}$$

The results are represented graphically in Figure 8(a) and Figure 8(b). From the Figure 8(a) and Figure 8(b), it is clear that quantum mechanical probability distribution is very different from the classical one. But when $n \to \infty$, then the quantum mechanical probability distribution becomes identical to the classical probability distribution.

10. Quantum Mechanical Tunelling and the Classical-Quantum Correspondence

Now we want to calculate the probability of finding the oscillator inside the classical turning points $(-a \leq x \leq a)$ and outside the classical turning point $(|x| > a)$. Classical turning points are the end points where velocity of the particle become zero. At these points kinetic energy is zero and the total energy is solely the potential energy. Beyond these classical turning points, potential energy is greater than the total energy and hence kinetic energy is negative. Then the velocity will be imaginary which is absurd. This indicates that the probability of finding the particle beyond these classical turning points is zero. But for the quantum linear SHO, probability distribution plots show that there is a finite probability of finding the particle outside the classical turning points.

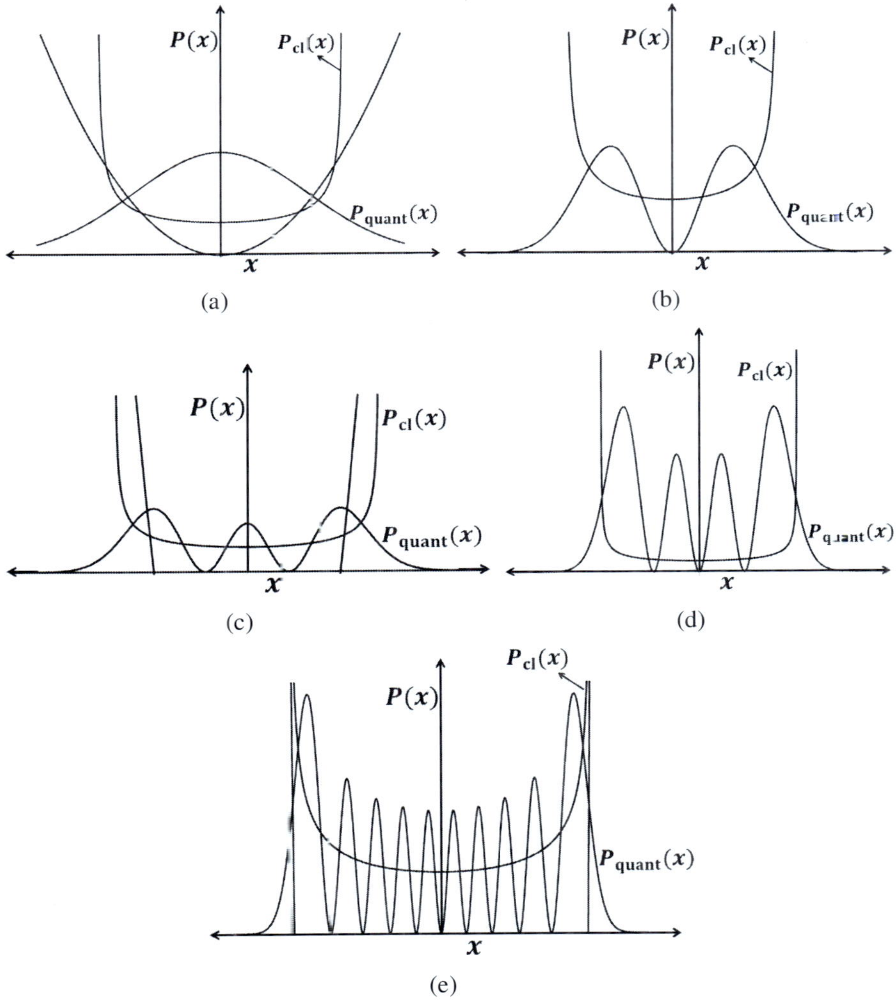

Figure 7. Probability distribution of linear Simple Harmonic Oscillator: classical and quantum.

This is a common phenomenon in quantum mechanics known as the quantum mechanical tunnelling [44]. It is a consequence of Heisenberg's uncertainty principle.

For a bound particle uncertainty in position (Δx) is finite. Therefore, the uncertainty in momentum (Δp_x) may have any higher value $\geq \dfrac{\hbar}{2\Delta x}$. If Δp_x is

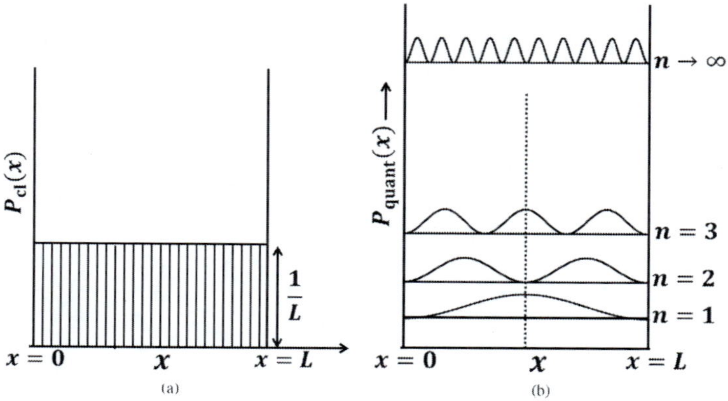

(a) (b)

Figure 8. (a) Probability distribution of classical free particle confined in a 1D box of length L. (b) Probability distribution of quantum free particle confined in a 1D box of length L.

arbitrary, then p_x as well as kinetic energy is arbitrary. Therefore, we cannot say with guarantee that total energy is less than the potential barrier. For this energy fluctuation, it may happen that in some instances total energy is actually greater then the potential barrier and it can jump over the potential barrier. Consequently, there is a finite probability of finding the particle even in the classically inaccessable region.

At the classical turning points, total energy is solely the potential energy and hence

$$V(a) = \frac{1}{2}m\omega^2 a^2 = E = \left(n + \frac{1}{2}\right)\hbar\omega \tag{88}$$

This gives

$$a = \pm\sqrt{\left(n + \frac{1}{2}\right)\frac{2\hbar}{m\omega}} = \pm\sqrt{\left(n + \frac{1}{2}\right)\frac{2\hbar}{\sqrt{mk}}} \tag{89}$$

Now for the ground state ($n = 0$), classical turning points $a = \pm\sqrt{\frac{\hbar}{\sqrt{mk}}}$ and the probability of finding the particle inside the classical turning points is

$$P_0(-a \leq x \leq a) = \int_{-a}^{a} \left(\frac{\alpha}{\pi}\right)^{\frac{1}{2}} e^{-\alpha x^2} dx = 2\int_{0}^{a} \left(\frac{\alpha}{\pi}\right)^{\frac{1}{2}} e^{-\alpha x^2} dx$$

Let $\sqrt{\alpha}\, x = y$, or, $dx = \frac{dy}{\sqrt{\alpha}}$ and when $x = 0, y = 0; x = a$ the $y = \sqrt{\alpha}\, a = 1$, as $\sqrt{\alpha}\, a = \left(\frac{\sqrt{k\mu}}{\hbar}\right)^{\frac{1}{2}} \times \left(\frac{\hbar}{\sqrt{mk}}\right)^{\frac{1}{2}} = 1$. Therefore,

$$P_0(-a \leq x \leq a) = 2\int_{0}^{1} \left(\frac{\alpha}{\pi}\right)^{\frac{1}{2}} e^{-y^2} \frac{dy}{\sqrt{\alpha}} = \frac{2}{\sqrt{\pi}}\int_{0}^{1} e^{-y^2} dy = \mathrm{erf}(1)$$

$$= 0.842701$$

$\mathrm{erf(z)} = \frac{2}{\sqrt{\pi}}\int_{0}^{z} e^{-x^2} dx$ is called the error function. Value of erf(1) is 0.842701. Hence, for the ground state of linear SHO, probability of finding the particle in between the classical turning point is 0.842701 and probability of finding the particle outside the classical turning point $= (1 - 0.842701)$ or 0.157299.

Let us now calculate the probability of finding the particle inside and outside the classical turning points for several excited state. We have derived the expressions for different states as follows.

For $n = 1$, $a = \pm\sqrt{3\frac{\hbar}{\sqrt{mk}}}$ and probability of finding the particle inside the classical turning points

$$\begin{aligned} P_1(-a \leq x \leq a) &= \int_{-a}^{a} 2\left(\frac{\alpha}{\pi}\right)^{\frac{1}{2}} \alpha x^2 e^{-\alpha x^2} dx \\ &= 2\int_{0}^{a} 2\left(\frac{\alpha}{\pi}\right)^{\frac{1}{2}} \alpha x^2 e^{-\alpha x^2} dx \\ &= \int_{0}^{\sqrt{3\frac{\hbar}{\sqrt{k\mu}}}} 2\left(\frac{\alpha}{\pi}\right)^{\frac{1}{2}} \alpha x^2 e^{-\alpha x^2} dx \\ &= -\frac{2\sqrt{3}}{\sqrt{\pi}} + \text{erf}\left(\sqrt{3}\right) \\ &= 0.88839 \end{aligned} \tag{90}$$

Probability of finding the particle outside the classical turning point is $(1 - 0.88839) = 0.11161$. For other states similar calculation can be done. These results are summarized in table (2) and Figure (9). From the plot it is

Table 2. Table for probability of finding the particle inside and outside the classical turning point for a linear SHO

Quan. no. (n)	$P(-a \leq x \leq a)$	$P(\mid x \mid > \mid a \mid)$
0	0.842701	0.157299
1	0.88839	0.11161
2	0.904931	0.0950694
3	0.909072	0.0909275
4	0.921074	0.0789264
5	0.925966	0.0740342
6	0.929819	0.0701809
7	0.932969	0.0670313
8	0.935614	0.064386

clear that the probability of finding the particle outside the classical turning points gradually decreases with increase in quantum number. From this trend it is again clear that as $n \to \infty$, probability of finding the particle outside the classical turning point will be zero. This is the results classically expected. This implies that as quantum number $n \to \infty$, quantum mechanical results will match with the classically expected results. This is a consequence of Bohr's correspondence principle.

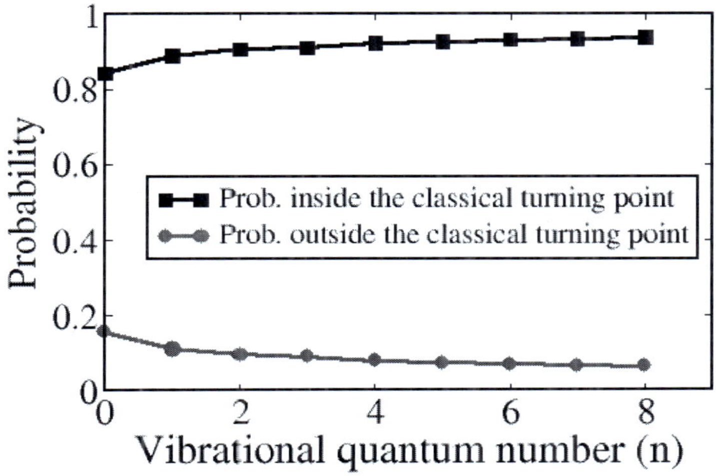

Figure 9. Probability of finding the particle inside and outside the classical turning points for a linear SHO.

11. Virial Theorem and Classical-Quantum Correspondence

For a system of N particles

$$2\langle T\rangle = -\sum_{k=1}^{N}\left(\vec{F_k}.\vec{r_k}\right) \tag{91}$$

where, $\vec{F_k}$ is the force of the k-th particle located at a point $\vec{r_k}$ and $\langle T\rangle$ is the average kinetic energy. This is the classical virial theorem [45]. If the force of a particle comes from a potential of the form $V(r) = ar^s$, then virial theorem takes the form

$$2\langle T\rangle = s\langle V\rangle \tag{92}$$

where, $\langle V\rangle$ is the average potential energy. This also implies that, if total energy of the system is E, i.e. $E = \langle T\rangle + \langle V\rangle$, then

$$\langle T\rangle = \frac{s}{s+2}E \tag{93}$$

and

$$\langle V \rangle = \frac{2}{s+2} E \tag{94}$$

Although originally derived for classical mechanics the virial theorem also holds for quantum mechanics and was first shown by Fock [43]. Let us check the virial theorem for classical linear SHO. We have already seen that for classical linear SHO, the probability of finding a particle within the region x and $(x + dx)$ is given by

$$P_{\mathrm{cl}}(x)dx = \frac{dx}{\pi\sqrt{a^2 - x^2}} \tag{95}$$

and velocity

$$v = \omega\sqrt{a^2 - x^2} \tag{96}$$

Now, $dv = -\dfrac{xdx}{\omega\sqrt{a^2 - x^2}}$, hence, $dx = -\dfrac{\omega\sqrt{a^2 - x^2}\, dv}{x}$. Using equation (96), we get

$$dx = -\frac{vdv}{\omega^2\sqrt{a^2 - \frac{v^2}{\omega^2}}} \tag{97}$$

Now the velocity distribution i.e. probability of finding particle having velocity within the range v and $v + dv$ is given by

$$P_{\mathrm{cl}}(v)dv = -\frac{dv}{\pi\sqrt{\omega^2 a^2 - v^2}} \tag{98}$$

With this velocity distribution we can evaluate the average kinetic energy as

$$\begin{aligned} \langle T \rangle &= \int_{-v_{\max}}^{v_{\max}} T \frac{dv}{\pi\sqrt{\omega^2 a^2 - v^2}} \\ &= 2\int_{0}^{v_{\max}} \frac{1}{2} m v^2 \frac{dv}{\pi\sqrt{\omega^2 a^2 - v^2}} \end{aligned} \tag{99}$$

Velocity is maximum at the equilibrium position i.e. at $x = 0$. This gives $v_{max} = a\omega$ and we get

$$\langle T \rangle = m \int_0^{a\omega} \frac{v^2 dv}{\pi\sqrt{\omega^2 a^2 - v^2}} \tag{100}$$

By substituting $v = a\omega \sin\theta$ and changing the integration limit 0 to $\frac{\pi}{2}$, we get the average kinetic energy as

$$\langle T \rangle = \frac{1}{4} m\omega^2 a^2 \tag{101}$$

Now we calculate the average potential energy $\langle V \rangle$ as follows.

$$\begin{aligned} \langle V \rangle &= \int_{-a}^{a} V(x) \frac{dx}{\pi\sqrt{a^2 - x^2}} \\ &= 2\int_0^a \frac{1}{2} m\omega^2 x^2 \frac{dx}{\pi\sqrt{a^2 - x^2}} \\ &= m\omega^2 \int_0^a \frac{x^2 dx}{\pi\sqrt{a^2 - x^2}} \end{aligned} \tag{102}$$

Again substituting $x = a\sin\theta$ and changing the limit 0 to $\frac{\pi}{2}$, We get,

$$\langle V \rangle = \frac{1}{4} m\omega^2 a^2 \tag{103}$$

From equations (101) and (103), we get

$$\langle T \rangle = \langle V \rangle = \frac{E}{2} \tag{104}$$

This satisfies the virial theorem equation (92) with $s = 2$ as for linear SHO $V(x) = \frac{1}{2} m\omega^2 x^2$.

Now let us now check the virial theorem for quantum linear SHO. For quantum linear SHO average kinetic and potential energy in an arbitrary normalized quantum state ψ_v are given by

$$\langle T\rangle = \langle\psi_n|\hat{T}|\psi_n\rangle = \frac{1}{2m}\langle\psi_n|\hat{p_x^2}|\psi_n\rangle = \frac{\langle p_x^2\rangle}{2m} \tag{105}$$

$$\text{and} \quad \langle V\rangle = \langle\psi_n|\hat{V}|\psi_n\rangle = \frac{1}{2}m\omega^2\langle\psi_n|\hat{x^2}|\psi_n\rangle = \frac{1}{2}m\omega^2\langle x^2\rangle \tag{106}$$

$\langle p_x^2\rangle$ and $\langle x^2\rangle$ can be easily calculated using ladder operators.

Using the relations given in section (53) and (54),

$$\begin{aligned}
\langle p_x^2\rangle = \langle\psi_n|\hat{p}_x^2|\psi_n\rangle &= \left\langle\psi_n\right|\left[-\frac{\hbar m\omega}{2}\left(\hat{a}_+ - \hat{a}_-\right)\left(\hat{a}_+ - \hat{a}_-\right)\right]\left|\psi_n\right\rangle \\
\text{or,} \quad \langle p_x^2\rangle &= -\frac{\hbar m\omega}{2}\langle\psi_n|\left(\hat{a}_+^2 - \hat{a}_+\hat{a}_- - \hat{a}_-\hat{a}_+ + \hat{a}_-^2\right)|\psi_n\rangle \\
&= -\frac{\hbar m\omega}{2}\left[\langle\psi_n|\hat{a}_+^2|\psi_n\rangle - \langle\psi_n|\hat{a}_+\hat{a}_-|\psi_n\rangle - \langle\psi_n|\hat{a}_-\hat{a}_+|\psi_n\rangle\right. \\
&\qquad \left.+\langle\psi_n|\hat{a}_-^2|\psi_n\rangle\right] \\
&= \frac{\hbar m\omega}{2}\left[\langle\psi_n|\hat{a}_+\sqrt{n}|\psi_{n-1}\rangle + \langle\psi_n|\hat{a}_-\sqrt{n+1}|\psi_{n+1}\rangle\right] \\
&= \frac{\hbar m\omega}{2}\left[\sqrt{n}\langle\psi_n|\sqrt{n}|\psi_n\rangle + \sqrt{n+1}\langle\psi_n|\sqrt{n+1}|\psi_n\rangle\right] \\
&= \frac{\hbar m\omega}{2}\left[n+n+1\right] \\
&= \left(n+\frac{1}{2}\right)\hbar m\omega
\end{aligned} \tag{107}$$

As $\langle\psi_n|\psi_n\rangle = 1$ from normalization relation. First and fourth terms are zero as $\hat{a}_+^2|\psi_n\rangle \propto |\psi_{n+2}\rangle$ and $\hat{a}_-^2|\psi_n\rangle \propto |\psi_{n-2}\rangle$ which are orthogonal to ψ_n. Therefore the average kinetic energy

$$\langle T\rangle = \frac{\langle p_x^2\rangle}{2m} = \frac{1}{2}\left(n+\frac{1}{2}\right)\hbar\omega \tag{108}$$

Similarly

$$\begin{aligned}
\langle x^2\rangle = \langle\psi_n|\hat{x}^2|\psi_n\rangle &= \left\langle\psi_n\right|\left[\frac{\hbar}{m\omega}\left(\hat{a}_+ + \hat{a}_-\right)\left(\hat{a}_+ + \hat{a}_-\right)\right]\left|\psi_n\right\rangle \\
\text{or,} \quad \langle x^2\rangle &= \left(n+\frac{1}{2}\right)\frac{\hbar}{m\omega}
\end{aligned} \tag{109}$$

This gives

$$\langle V\rangle = \frac{1}{2}m\omega^2\langle x^2\rangle = \frac{1}{2}\left(v+\frac{1}{2}\right)\hbar\omega \tag{110}$$

From equations (108) and (110), it is clear that the virial equation is satisfied for the quantum linear SHO.

Validity of virial theorem can also be checked for hydrogen atom, particle in a box problem and all other systems having power law potential. For H-atom problem potential energy

$$V(r) = -\frac{ze^2}{r} \propto r^{-1}$$

Hence, from virial theorem,

$$2\langle T\rangle = -\langle V\rangle \tag{111}$$

$$\text{and} \quad \langle E\rangle = \langle T\rangle = \frac{1}{2}\langle V\rangle \tag{112}$$

For a free particle of mass m confined in a box of length if potential energy is given as

$$\begin{aligned} V(x) &= 0 \text{ when } -\frac{L}{2} \leq x \leq \frac{L}{2} \\ &= \infty \text{ otherwise} \end{aligned}$$

Then this potential can also be written as

$$V(x) = \lim_{N\to\infty} \left|\frac{2x}{L}\right|^N \tag{113}$$

In this case, $s \to \infty$,

$$\langle T\rangle = \frac{s}{s+2}\langle E\rangle \tag{114}$$

implies

$$\begin{aligned} \langle T\rangle &= \langle E\rangle \\ \text{and} \quad \langle V\rangle &= \frac{2}{s+2}\langle E\rangle \end{aligned} \tag{115}$$

which implies $\langle V\rangle \to 0$.

If we consider an anharmonic oscillator with the potential energy $V(x) = ax^4$ then according to virial theorem,

$$2\langle T\rangle = 4\langle V\rangle \tag{116}$$

$$\langle T\rangle = \frac{2}{3}\langle E\rangle \tag{117}$$

$$\text{and} \quad \langle V\rangle = \frac{1}{3}\langle E\rangle \tag{118}$$

12. HELLMANN-FEYNMAN THEOREM AND CLASSICAL-QUANTUM CORRESPONDENCE

Hellmann-Feynman theorem [46–48] relates the derivative of total energy with respect to a parameter to the average value of the derivative of Hamiltonian operator with respect to the same parameter. Mathematically,

$$\frac{\partial E_\lambda}{\partial \lambda} = \frac{\left\langle \psi_\lambda \left| \frac{\partial \hat{H}_\lambda}{\partial \lambda} \right| \psi_\lambda \right\rangle}{\langle \psi_\lambda | \psi_\lambda \rangle} \tag{119}$$

where λ is the parameter on which both E and $\hat{H}$ depend. For normalized function the denominator can be taken as unity. For the linaer SHO, Hamiltonian operator is given as

$$\hat{H} = -\frac{\hbar^2}{2m}\frac{d^2}{dx^2} + \frac{1}{2}kx^2$$

If we consider the force constant k as the parameter,

$$\frac{\partial \hat{H}}{\partial k} = \frac{1}{2}x^2$$

Now,

$$\begin{aligned}
\left\langle \psi_n \left| \frac{\partial \hat{H}}{\partial k} \right| \psi_n \right\rangle &= \left\langle \psi_n \left| \frac{1}{2}x^2 \right| \psi_n \right\rangle = \frac{1}{k}\left\langle \psi_n \left| \frac{1}{2}kx^2 \right| \psi_n \right\rangle \\
&= \frac{1}{k}\langle V \rangle = \frac{1}{2k}\left(n + \frac{1}{2}\right)\hbar\omega \text{ [From Virial theorem]} \\
&= \frac{1}{2}\left(n + \frac{1}{2}\right)\frac{\hbar}{m\omega}
\end{aligned} \tag{120}$$

Again, we know energy of a linear SHO is given as

$$E_n = \left(n + \frac{1}{2}\right)\hbar\omega = \left(n + \frac{1}{2}\right)\hbar\sqrt{\frac{k}{m}}$$

Now

$$\begin{aligned}
\frac{\partial E_n}{\partial k} = \left(n + \frac{1}{2}\right)\frac{\hbar}{\sqrt{m}}\frac{1}{2\sqrt{k}} &= \frac{1}{2}\left(n + \frac{1}{2}\right)\frac{\hbar}{\sqrt{mk}} \\
&= \frac{1}{2}\left(n + \frac{1}{2}\right)\frac{\hbar}{m\omega}
\end{aligned} \tag{121}$$

From equations (120) and (121), it is clear that for a linear SHO, Hellmann-Feynman theorem is applicable.

13. Classical and Quantum Dynamics and the Correspondence Principle

The harmonic oscillator wave mechanics agrees with ordinary mechanics had been shown by Schrödinger in the article under the title "The continuous transition from micro- to macro-mechanics" [49]. A general and direct line of connection between quantum mechanics and Newtonian mechanics was established in 1927 by Ehrenfest [38–41], who showed by a short elementary calculation without approximations, that the average value of the time derivative of the momentum is equal to the expectation value of the negative gradient of the potential energy function. Ehrenfest's affirmation of Newton's second law in the sense of average taken over the wave packet had a great appeal to many physicists. Due to this, it is possible to describe the particle by a localized wave packet which, eventually spreading out in space, follows the trajectory of the classical motion.

The time evolution of the average value of a classical observable in a normalized quantum state ψ can be written as

$$\begin{aligned}
\frac{\partial\langle A\rangle}{\partial t} = \frac{\partial}{\partial t}\langle\psi\left|\hat{A}\right|\psi\rangle &= \left\langle\frac{\partial\psi}{\partial t}\left|\hat{A}\right|\psi\right\rangle + \left\langle\psi\left|\frac{\partial\hat{A}}{\partial t}\right|\psi\right\rangle + \left\langle\psi\left|\hat{A}\right|\frac{\partial\psi}{\partial t}\right\rangle \\
&= \left\langle\frac{\hat{H}\psi}{i\hbar}\left|\hat{A}\right|\psi\right\rangle + \left\langle\frac{\partial A}{\partial t}\right\rangle + \left\langle\psi\left|\hat{A}\right|\frac{\hat{H}\psi}{i\hbar}\right\rangle \\
&= \frac{1}{i\hbar}\left[\left\langle\psi|\hat{A}\hat{H}|\psi\right\rangle - \left\langle\psi|\hat{H}\hat{A}|\psi\right\rangle\right] + \left\langle\frac{\partial A}{\partial t}\right\rangle \\
&= \frac{1}{i\hbar}\left\langle\psi\left|\left(\hat{A}\hat{H} - \hat{H}\hat{A}\right)\right|\right\rangle + \left\langle\frac{\partial A}{\partial t}\right\rangle \\
&= \frac{\left\langle\left[\hat{A},\hat{H}\right]\right\rangle}{i\hbar} + \left\langle\frac{\partial A}{\partial t}\right\rangle
\end{aligned} \tag{122}$$

This is the Heisenberg's equation of motion. If A does not depend on time explicitly then

$$\frac{\partial\langle A\rangle}{\partial t} = \frac{\left\langle\left[\hat{A},\hat{H}\right]\right\rangle}{i\hbar} \tag{123}$$

If we take $A \equiv x$ in equation (123) then,

$$\frac{\partial\langle x\rangle}{\partial t} = \frac{1}{i\hbar}\left\langle\left[\hat{x},\hat{H}\right]\right\rangle \tag{124}$$

In one dimension

$$\begin{aligned}\left[\hat{x},\hat{H}\right]=\left[\hat{x},\frac{\hat{p}_x^2}{2m}+\hat{V}\right] &= \frac{1}{2m}\left[\hat{x},\hat{p}_x^2\right]+\left[\hat{x},\hat{V}\right]\\ &= \frac{1}{2m}\times 2i\hbar\hat{p}_x+0\\ \text{or,}\quad \left[\hat{x},\hat{H}\right] &= \frac{i\hbar\hat{p}_x}{m}\end{aligned} \tag{125}$$

Hence,

$$\frac{\partial\langle x\rangle}{\partial t}=\frac{1}{i\hbar}\left\langle i\hbar\frac{p_x}{m}\right\rangle = \frac{\langle p_x\rangle}{m} \tag{126}$$

Again if we write $A \equiv p_x$ in equation (123), then

$$\frac{\partial p_x}{\partial t} = \left\langle\left[\hat{p}_x,\hat{H}\right]\right\rangle \tag{127}$$

In one dimension,

$$\left[\hat{p}_x,\hat{H}\right]=\left[\hat{p}_x,\frac{\hat{p}_x^2}{2m}+\hat{V}\right] = 0+\left[\hat{p}_x,\hat{V}\right]=-i\hbar\frac{\partial V(x)}{\partial x} \tag{128}$$

Therefore,

$$\frac{\partial\langle p_x\rangle}{\partial t} = \frac{1}{i\hbar}\left\langle\left(-i\hbar\frac{\partial V(x)}{\partial x}\right)\right\rangle=-\left\langle\frac{\partial V(x)}{\partial x}\right\rangle \tag{129}$$

Therefore, in one dimension we get

$$\frac{d}{dt}\langle x\rangle = \frac{\langle p_x\rangle}{m} \tag{130}$$

$$\text{and}\quad \frac{d}{dt}\langle p_x\rangle = -\left\langle\frac{\partial V(x)}{\partial x}\right\rangle \tag{131}$$

These are the Ehrenfest equation of motion. These equation are jointly reminiscent to the first-order cannonical equations of motion

$$\frac{dx}{dt} = \frac{p_x}{m} \tag{132}$$

$$\text{and}\quad \frac{dp_x}{dt} = -\frac{\partial V(x)}{\partial x} \tag{133}$$

which are associated with the classical systems of type $H(x,p) = \frac{p^2}{2m} + V(x)$.

To show the analogy of classical and quantum equations of motion we start from the Newton's equation of motion for linear SHO and it is given by

$$\frac{d^2x}{dt^2} + \omega^2 x = 0 \tag{134}$$

Here x is a function of time. Now to investigate the dynamics of x^2, we write

$$\frac{d}{dt}x^2 = 2x\,\frac{dx}{dt} \tag{135}$$

$$\text{and} \quad \frac{d^2}{dt^2}x^2 = \frac{d}{dt}\left(2x\frac{dx}{dt}\right) = 2x\frac{d^2x}{dt^2} + 2\left(\frac{dx}{dt}\right)^2$$

$$\text{or,} \quad \frac{d^2}{dt^2}x^2 + 2\omega^2x^2 = \frac{2}{m^2}p_x^2 \tag{136}$$

Let us investigate the dynamics of p_x and p_x^2.

$$\frac{dp_x}{dt} = F = -\frac{dV(x)}{dx} = -\frac{d}{dx}\left(\frac{1}{2}m\omega^2x^2\right) = -m\omega^2 x$$

$$\text{and} \quad \frac{d^2p_x}{dt^2} = \frac{d}{dt}\left(-m\omega^2x\right) = -m\omega^2\frac{dx}{dt} = -\omega^2 p_x$$

$$\text{or,} \quad \frac{d^2p_x}{dp^2} + \omega^2 p_x = 0$$

Now,

$$\frac{dp_x^2}{dt} = 2p_x\frac{dp_x}{dt}$$

$$\text{or,} \quad \frac{d^2}{dt^2}p_x^2 = \frac{d}{dt}\left(2p_x\frac{dp_x}{dt}\right)$$

$$= 2p_x\frac{d^2p_x}{dt^2} + 2\left(\frac{dp_x}{dt}\right)^2$$

$$= 2p_x \times \left(-\omega^2p_x\right) + 2\times\left(-m\omega^2x\right)^2$$

$$\text{or,} \quad \frac{d^2}{dt^2}p_x^2 + 2\omega^2p_x^2 = 2m^2\omega^4x^4 \tag{137}$$

Here a point to note that equation (136) and equation (137) are coupled equations. The solution scheme will be presented in the context of quantum dynamics.

Now to find the dynamics of classical fluctuation we write

$$\begin{aligned}\frac{d}{dt}(\Delta x)^2 &= \frac{d}{dt}\left(\langle x^2\rangle - \langle x\rangle^2\right) = \frac{d}{dt}\langle x^2\rangle \text{ as } \langle x\rangle = 0\\ \text{and } \frac{d}{dt}(\Delta p_x^2) &= \frac{d}{dt}\left[\langle p_x^2\rangle - \langle p_x\rangle^2\right] = \frac{d}{dt}\langle p_x^2\rangle \text{ as } \langle p_x\rangle = 0\end{aligned}$$

Hence,

$$\begin{aligned}\frac{d}{dt}\left[\frac{\Delta p_x^2}{2m} + \frac{1}{2}m\omega^2(\Delta x)^2\right] &= \frac{1}{2m}\frac{d}{dt}\langle p_x^2\rangle + \frac{1}{2}m\omega^2\frac{d}{dt}\langle x^2\rangle\\ &= \frac{d}{dt}\left[\frac{\langle p_x^2\rangle}{2m} + \frac{1}{2}m\omega^2\langle x^2\rangle\right]\\ &= \frac{d}{dt}\langle E\rangle = 0 \text{ as it is a conservative system}\end{aligned}$$

In quantum mchanics Newton's equations of motion cannot be used as it is. This is due to uncertainy principle. Position and momentum cannot be determined simultaneously with arbitrary precision. But it is the Ehrenfest equation of motion that gives a way out. This gives equations of motions similar to Newton's equation of motion but the instanteneous positions and momentum should be replaced by their expectation values.

For linear SHO,

$$\begin{aligned}\frac{d}{dt}\langle x\rangle = \frac{\left\langle[\hat{x},\hat{H}]\right\rangle}{i\hbar} &= \frac{1}{i\hbar}\left\langle\left[\hat{x}, \frac{\hat{p_x^2}}{2m} + \frac{1}{2}m\omega^2\hat{x^2}\right]\right\rangle\\ &= \frac{1}{i\hbar}\left\langle\left[\hat{x}, \frac{\hat{p_x^2}}{2m}\right] + \frac{1}{2}m\omega^2\left[\hat{x},\hat{x^2}\right]\right\rangle\\ &= \frac{1}{2mi\hbar}\langle 2i\hbar\hat{p_x}\rangle\\ \text{or, } \frac{d}{dt}\langle x\rangle &= \frac{\langle p_x\rangle}{m} \qquad (138)\end{aligned}$$

and

$$\begin{aligned}
\frac{d}{dt}\langle p_x\rangle = \frac{1}{i\hbar}\left\langle[\hat{p_x},\hat{H}]\right\rangle &= \frac{1}{i\hbar}\left\langle\left[\hat{p_x},\frac{\hat{p_x^2}}{2m}+\frac{1}{2}m\omega^2\hat{x^2}\right]\right\rangle \\
&= \frac{1}{i\hbar}\left\langle\frac{1}{2m}\left[p_x,\hat{p_x^2}\right]+\frac{1}{2}m\omega^2\left[\hat{p_x},\hat{x^2}\right]\right\rangle \\
&= \frac{1}{i\hbar}\left\langle\frac{1}{2}m\omega^2\left[\hat{p_x},\hat{x^2}\right]\right\rangle \\
&= \frac{1}{i\hbar}\frac{1}{2}m\omega^2(-2i\hbar)\langle x\rangle \\
\text{or,}\quad \frac{d}{dt}\langle p_x\rangle &= -m\omega^2\langle x\rangle \qquad (139)
\end{aligned}$$

To find the dynamics of $\langle x\rangle$ we write,

$$\begin{aligned}
\frac{d^2}{dt^2}\langle x\rangle = \frac{d}{dt}\left(\frac{d}{dt}\langle x\rangle\right) &= \frac{d}{dt}\frac{\langle p_x\rangle}{m} = \frac{1}{m}\left(-m\omega^2\langle x\rangle\right) \\
\text{or,}\quad \frac{d^2}{dt^2}\langle x\rangle &= -\omega^2\langle x\rangle \qquad (140) \\
\text{or,}\quad \frac{d^2}{dt^2}\langle x\rangle + \omega^2\langle x\rangle &= 0 \qquad (141)
\end{aligned}$$

This equation matches exactly with the classical equation of motion given in equation (134). Now the dynamics of expectation value of momentum can be obtained as

$$\begin{aligned}
\frac{d^2}{dt^2}\langle p_x\rangle = \frac{d}{dt}\left(\frac{d}{dt}\langle p_x\rangle\right) &= \frac{d}{dt}\left(-m\omega^2\langle x\rangle\right) = -m\omega^2\frac{d}{dt}\langle x\rangle \\
\text{or,}\quad \frac{d^2}{dt^2}\langle p_x\rangle &= -m\omega^2\frac{\langle p_x\rangle}{m} \qquad (142) \\
\text{or,}\quad \frac{d^2}{dt^2}\langle p_x\rangle + \omega^2\langle p_x\rangle &= 0 \qquad (143)
\end{aligned}$$

This is similar to the classical dynamics of p_x. Now let us consider the dynamics of $\langle x^2\rangle$ and $\langle {p_x}^2\rangle$. From Heisenberg's equation of motion,

$$\begin{aligned}
\frac{d}{dt}\langle x^2\rangle = \frac{1}{i\hbar}\langle[\hat{x^2},\hat{H}]\rangle &= \frac{1}{i\hbar}\left\langle\left[\hat{x^2},\frac{\hat{{p_x}^2}}{2m}+\frac{1}{2}m\omega^2\hat{x^2}\right]\right\rangle\\
&= \frac{1}{2mi\hbar}\left\langle\left[\hat{x^2},\hat{p_x^2}\right]\right\rangle\\
&= \frac{1}{2mi\hbar}\left\langle\left[\hat{x^2},\hat{p_x}\right].\hat{p_x}+\hat{p_x}\left[\hat{x^2},\hat{p_x}\right]\right\rangle\\
&= \frac{1}{2mi\hbar}\langle 2i\hbar x.p_x + 2i\hbar p_x.x\rangle\\
\text{or,}\quad \frac{d}{dt}\langle x^2\rangle &= \frac{1}{m}\langle(x.p_x+p_x.x)\rangle \qquad (144)
\end{aligned}$$

and

$$\begin{aligned}
\frac{d}{dt}\langle p_x^2\rangle = \frac{1}{i\hbar}\left\langle\left[\hat{p_x^2},\hat{H}\right]\right\rangle &= \frac{1}{i\hbar}\left\langle\left[\hat{p_x^2},\frac{\hat{{p_x}^2}}{2m}+\frac{1}{2}m\omega^2\hat{x^2}\right]\right\rangle\\
&= \frac{1}{i\hbar}\times\frac{1}{2}m\omega^2\left\langle\left[\hat{{p_x}^2},\hat{x^2}\right]\right\rangle\\
&= \frac{1}{i\hbar}\times\frac{1}{2}m\omega^2\left\langle\left[\hat{{p_x}^2},\hat{x}.\hat{x}\right]\right\rangle\\
&= \frac{1}{i\hbar}\times\frac{1}{2}m\omega^2\left\langle\left[\hat{{p_x}^2},\hat{x}\right].\hat{x}+\hat{x}.\left[\hat{{p_x}^2}.\hat{x}\right]\right\rangle\\
&= \frac{1}{i\hbar}\times\frac{1}{2}m\omega^2\langle -2i\hbar p_x.x - 2i\hbar x.p_x\rangle\\
\text{or,}\quad \frac{d}{dt}\langle p_x^2\rangle &= -m\omega^2\ \langle(x.p_x+p_x.x)\rangle \qquad (145)
\end{aligned}$$

Now let us consider the time evolution of energy.

$$\begin{aligned}
\frac{d\langle E\rangle}{dt} &= \frac{d}{dt}\left\langle\frac{\hat{{p_x}^2}}{2m}+\frac{1}{2}m\omega^2\hat{x^2}\right\rangle = \frac{1}{2m}\frac{d}{dt}\langle p_x^2\rangle + \frac{1}{2}m\omega^2\frac{d}{dt}\langle x^2\rangle\\
&= -\frac{\omega^2}{2}\langle(x.p_x+p_x.x)\rangle + \frac{\omega^2}{2}\langle(x.p_x+p_x.x)\rangle\\
\text{or,}\quad \frac{d\langle E\rangle}{dt} &= 0 \qquad (146)
\end{aligned}$$

This equation clearly indicates that the total average energy is conserved in case of linear SHO. This is the same result as expected classically as it it a conservative system.

To evaluate the dynamical equation of second moment we write,

$$\frac{d^2}{dt^2}\langle x^2\rangle = \frac{d}{dt}\left(\frac{d}{dt}\langle x^2\rangle\right) = \frac{1}{m}\frac{d}{dt}\langle (x.p_x + p_x.x)\rangle \tag{147}$$

where

$$\begin{aligned}
\frac{d}{dt} & \langle (x.p_x + p_x.x)\rangle = \frac{1}{i\hbar}\left\langle\left[(\hat{x}\hat{p_x} + \hat{p_x}\hat{x}), \hat{H}\right]\right\rangle \\
= & \frac{1}{i\hbar}\left\langle\left[(\hat{x}\hat{p_x} + \hat{p_x}\hat{x}), \frac{\hat{p_x}^2}{2m} + \frac{1}{2}m\omega^2\hat{x}^2\right]\right\rangle \\
= & \frac{1}{i\hbar}\left\langle\left[\hat{x}\hat{p_x}, \frac{\hat{p_x}^2}{2m}\right] + \left[\hat{x}\hat{p_x}, \frac{1}{2}m\omega^2\hat{x}^2\right] + \left[\hat{p_x}\hat{x}, \frac{\hat{p_x}^2}{2m}\right] + \left[\hat{p_x}x, \frac{1}{2}m\omega^2\hat{x}^2\right]\right\rangle \\
= & \frac{1}{i\hbar}\left[\frac{1}{2m}\left\langle\left\{\hat{x}.\left[\hat{p_x}, \hat{p_x}^2\right] + \left[\hat{x}, \hat{p_x}^2\right]\hat{p_x}\right\}\right\rangle + \frac{1}{2}m\omega^2\left\langle\left\{\hat{x}\left[\hat{p_x}, \hat{x}^2\right] + \left[\hat{x}, \hat{x}^2\right]\hat{p_x}\right\}\right\rangle\right. \\
+ & \left.\frac{1}{2m}\left\langle\left\{\hat{p_x}\left[\hat{x}, \hat{p_x}^2\right] + \left[\hat{p_x}, \hat{p_x}^2\right]\hat{x}\right\}\right\rangle + \frac{1}{2}m\omega^2\left\langle\left\{\hat{p_x}\left[\hat{x}, \hat{x}^2\right] + \left[\hat{p_x}, \hat{x}^2\right]\hat{x}\right\}\right\rangle\right] \\
= & \frac{1}{i\hbar}\left[\frac{1}{2m}\langle 2i\hbar p_x^2\rangle + \frac{1}{2}m\omega^2\langle -2i\hbar x^2\rangle + \frac{1}{2m}\langle 2i\hbar p_x^2\rangle + \frac{1}{2}m\omega^2\langle -2i\hbar x^2\rangle\right] \\
= & 2\frac{\langle p_x^2\rangle}{m} - 2m\omega^2\langle x^2\rangle
\end{aligned} \tag{148}$$

Hence from equations (147) and (148),

$$\begin{aligned}
\frac{d^2}{dt^2}\langle x^2\rangle &= 2\frac{\langle p_x^2\rangle}{m^2} - 2\omega^2\langle x^2\rangle \\
\text{or,} \quad \frac{d^2}{dt^2}\langle x^2\rangle + 2\omega^2 x^2 &= \frac{2}{m^2}\langle p_x^2\rangle
\end{aligned} \tag{149}$$

This equation is similar to the classical dynamical equation of x^2 as given in equation (136).

Let us now evaluate the quantum dynamical equation of $\langle p_x^2\rangle$ using equations (145) and (148) we get

$$\begin{aligned}
\frac{d^2}{dt^2}\langle p_x^2\rangle = \frac{d}{dt}\left(\frac{d}{dt}\langle p_x^2\rangle\right) &= -m\omega^2\frac{d}{dt}\langle (x.p_x + p_x.x)\rangle \\
\text{or,} \quad \frac{d^2}{dt^2}\langle p_x^2\rangle &= -2\omega^2\langle p_x^2\rangle + 2m^2\omega^4\langle x^2\rangle \\
\text{or,} \quad \frac{d^2}{dt^2}\langle p_x^2\rangle + 2\omega^2\langle p_x^2\rangle &= 2m^2\omega^4\left(x^4\right)
\end{aligned} \tag{150}$$

This is exactly the similar equation to that of equation (137). Here a point to note that equation (149) and (150) are coupled differential equations. These equations can be solved as follows.

Multiplying equation (149) by $\frac{1}{2}m\omega^2$ and equation (150) by $\frac{1}{2m}$ and adding them, we get

$$\frac{d^2}{dt^2}\left[\frac{1}{2m}\langle {p_x}^2\rangle + \frac{1}{2}m\omega^2\langle x^2\rangle\right] = 0 \tag{151}$$

After integration we get,

$$\frac{d}{dt}\left[\frac{1}{2m}\langle {p_x}^2\rangle + \frac{1}{2}m\omega^2\langle x^2\rangle\right] = C \tag{152}$$

where C is a constant of integration. It is clear that $\left[\frac{1}{2m}\langle {p_x}^2\rangle + \frac{1}{2}m\omega^2\langle x^2\rangle\right]$ represents the total energy and it is constant for linear SHO as expected for a conservative system. Hence C must be zero. Therefore

$$\frac{d}{dt}\left[\frac{1}{2m}\langle {p_x}^2\rangle + \frac{1}{2}m\omega^2\langle x^2\rangle\right] = 0 \tag{153}$$

Integrating again we get

$$\frac{1}{2m}\langle {p_x}^2\rangle + \frac{1}{2}m\omega^2\langle x^2\rangle = D \tag{154}$$

Here a point to mention that this constant D is nothing but the expectation value of energy of the system. Then

$$\langle {p_x}^2\rangle = 2mD - m^2\omega^2\langle x^2\rangle \tag{155}$$

Putting this value in equation (149) we get,

$$\frac{d^2}{dt^2}\langle x^2\rangle + 4\omega^2\langle x^2\rangle = \frac{4D}{m} \tag{156}$$

$$\text{or,}\quad \frac{d^2\langle x^2\rangle}{dt^2} + \Omega_1^2\langle x^2\rangle = \frac{4D}{m} \quad \text{where } \Omega_1 = 2\omega \tag{157}$$

From the equation (154) we get

$$\langle x^2\rangle = \frac{2D}{m\omega^2} - \frac{\langle {p_x}^2\rangle}{m^2\omega^2} \tag{158}$$

Putting this value of $\langle x^2 \rangle$ in equation (150) we get,

$$\frac{d^2}{dt^2}\langle {p_x}^2 \rangle + 4\omega^2 \langle {p_x}^2 \rangle = 4m\omega^2 D \tag{159}$$

$$\text{or,} \quad \frac{d^2}{dt^2}\langle p_x^2 \rangle + \Omega_1^2 \langle p_x^2 \rangle = 4m\omega^2 D \tag{160}$$

Equations (156) and (159) are decoupled and hence they can be solved for a fixed average energy and we can get trajectory.

13.1. Fluctuation Dynamics

Now let us concentrate on the dynamics of uncertianty in position and momentum, i.e. the fluctuation dynamics.

$$\begin{aligned}\frac{d}{dt}\left(\Delta x^2\right) &= \frac{d}{dt}\left[\langle x^2 \rangle - \langle x \rangle^2\right] \\ &= \left[\frac{d}{dt}\langle x^2 \rangle - 2\langle x \rangle \frac{d}{dt}\langle x \rangle\right]\end{aligned} \tag{161}$$

Using equation (144) and remembering that $\langle x \rangle = 0$, and using equation (144) we get

$$\frac{d}{dt}\left(\Delta x^2\right) = \frac{1}{m}\langle (x.p_x + p_x.x) \rangle \tag{162}$$

Again

$$\begin{aligned}\frac{d}{dt}\left(\Delta p_x^2\right) &= \frac{d}{dt}\left[\langle p_x^2 \rangle - \langle p_x \rangle^2\right] \\ &= \left[\frac{d}{dt}\langle p_x^2 \rangle - 2\langle p_x \rangle \frac{d}{dt}\langle p_x \rangle\right]\end{aligned} \tag{163}$$

Using equation (145) and remembering that $\langle p_x \rangle = 0$ and using equation (145) we get

$$\frac{d}{dt}\left(\Delta p_x^2\right) = -m\omega^2 \langle (x.p_x + p_x.x) \rangle \tag{164}$$

Hence,

$$\begin{aligned}\frac{1}{2m}\frac{d}{dt}\left(\Delta {p_x}^2\right)+\frac{1}{2}m\omega^2\frac{d}{dt}\left(\Delta x^2\right) &= -\frac{1}{2m}\left[m\omega^2\langle(x.p_x+p_x.x)\rangle\right]\\ &\quad+\frac{1}{2}m\omega^2\left[\frac{1}{m}\langle(x.p_x+p_x.x)\rangle\right]\\ \text{or, } \frac{d}{dt}\left[\frac{(\Delta {p_x}^2)}{2m}+\frac{1}{2}m\omega^2\left(\Delta x^2\right)\right] &= 0. \qquad (165)\end{aligned}$$

This means $\left[\frac{(\Delta {p_x}^2)}{2m}+\frac{1}{2}m\omega^2\left(\Delta x^2\right)\right]$ is constant. Hence it is clear that not only energy but also fluctuations are constants both for classical as well as quantum linear SHO.

Now

$$\begin{aligned}\frac{d^2}{dt^2}\left(\Delta x^2\right) &= \frac{d^2}{dt^2}\left[\langle x^2\rangle-\langle x\rangle^2\right]\\ &= \frac{d^2}{dt^2}\langle x^2\rangle-\frac{d^2}{dt^2}\langle x\rangle^2\\ &= \frac{d^2}{dt^2}\langle x^2\rangle-\frac{d}{dt}\left[2\langle x\rangle\frac{d\langle x\rangle}{dt}\right]\\ \text{or, } \frac{d^2}{dt^2}\left(\Delta x^2\right) &= \frac{d^2}{dt^2}\langle x^2\rangle-2\langle x\rangle\frac{d^2}{dt^2}\langle x\rangle-2\left(\frac{d\langle x\rangle}{dt}\right)^2 \qquad (166)\end{aligned}$$

Using equations (138), (149) and remembering that $\langle x\rangle=0$, we get

$$\begin{aligned}\frac{d^2}{dt^2}\left(\Delta x^2\right) &= 2\frac{\langle p_x^2\rangle}{m^2}-2\omega^2\langle x^2\rangle \qquad (167)\\ \text{or, } \frac{d^2}{dt^2}\left(\Delta x^2\right)+2\omega^2\langle x^2\rangle &= 2\frac{\langle {p_x}^2\rangle}{m^2}\\ \text{or, } \frac{d^2}{dt^2}\left(\Delta x^2\right)+2\omega^2\left(\Delta x^2\right) &= 2\frac{\langle {p_x}^2\rangle}{m^2} \text{ as } \langle x\rangle=0\end{aligned}$$

Again using virial theorem we get $\langle p_x^2\rangle = mE = mD$. Putting this value we get,

$$\begin{aligned}\frac{d^2}{dt^2}\left(\Delta x^2\right)+2\omega^2\left(\Delta x^2\right) &= \frac{2D}{m} \qquad (168)\\ \text{or, } \frac{d^2}{dt^2}\left(\Delta x^2\right)+\Omega_2^2\left(\Delta x^2\right) &= \frac{2D}{m} \text{ where } \Omega_2=\sqrt{2}\omega\end{aligned}$$

This equation is similar to equation (156). But there is a difference in angular velocity. x^2 evolve with time with an angular velocity $\Omega_1 = 2\omega$ where as Δx^2 evolves with time with an angular velocity $\Omega_2 = \sqrt{2}\omega$.

Similarly,

$$\begin{aligned}
\frac{d^2}{dt^2}\left(\Delta p_x^2\right) &= \frac{d^2}{dt^2}\left[\langle p_x^2\rangle - \langle p_x\rangle^2\right] \\
&= \frac{d^2}{dt^2}\langle p_x^2\rangle - \frac{d^2}{dt^2}\langle p_x\rangle^2 \\
&= \frac{d^2}{dt^2}\langle p_x^2\rangle - \frac{d}{dt}\left[2\langle p_x\rangle\frac{d}{dt}\langle p_x\rangle\right] \\
&= \frac{d^2}{dt^2}\langle p_x^2\rangle - 2\left(\frac{d}{dt}\langle p_x\rangle\right)^2 \qquad (169)
\end{aligned}$$

Using equations (139),(150) and as $\langle x\rangle = 0$ we get

$$\begin{aligned}
\frac{d^2}{dt^2}\left(\Delta p_x^2\right) &= -2\omega^2\langle p_x^2\rangle + 2\omega^4 m^2\langle x^2\rangle - 2m^2\omega^4\langle x\rangle^2 \\
&= -2\omega^2\langle p_x^2\rangle + 2\omega^4 m^2\langle x^2\rangle \\
&= 4m\omega^2\left[\frac{\langle p_x^2\rangle}{2m} + \frac{1}{2}m\omega^2\langle x^2\rangle\right] - 4\omega^2\langle p_x^2\rangle \\
&= 4m\omega^2 D - 4\omega^2\langle p_x^2\rangle \qquad (170)
\end{aligned}$$

$$\text{or,}\quad \frac{d^2}{dt^2}\left(\Delta p_x^2\right) + 4\omega^2\left(\Delta p_x^2\right) = 4m\omega^2 D \qquad (171)$$

$$\text{or,}\quad \frac{d^2}{dt^2}\left(\Delta p_x^2\right) + \Omega_1^2\left(\Delta p_x^2\right) = 4m\omega^2 D \qquad (172)$$

This is similar to equation (159). p_x^2 as well as Δp_x^2 both evolves with time with the same angular velocity 2ω.

CONCLUSION

Although laws of classical mechanics can explain more or less all the essential physics in the macroscopic world, it fails in the microscpic or subatomic world. Where as quantum mechanics is general and can be applicable for both microscopic and macroscopic world. Critical investigation reveals that there is classical-quantum correspondence. Quantum mechanically calculated results match with the classically expected results when quantum number is very high. This is the Bohr's correspondence principle. This central principle has been illustrated taking linear SHO as an example. SHO is one of the most important systems for several reasons. Any stable molecular potential can be approximated by SHO near the equilibrium point. We have calculated probability distribution, quantum mechanical tunneling, classical and quantum dynamics, virial theorems, Hellmann-Feynmann theorem etc. We noticed that all these quantum mechanical results finally tend to classical results in the high quantum number limit and hence justified the correspondence principle.

Acknowledgment

We sincerely thank Professor Jayanta Kumar Bhattacharjee, IACS for a lot of stimulating discussions and critical reading of the manuscript. One of us (PG) sincerely acknowledges IACS,Kolkata for her research fellowship.

REFERENCES

[1] Goldstein, H., Poole, C. P. and Safko, J. L. (2001). *Classical Mechanics*, Third Edition, Addison-Wesley.

[2] Landau, L. D. and Lifshitz, E. M. (1976). *Course of Theoretical Physics, Mechanics*. Butterworth-Heinemann, 1, Third Edition.

[3] Landau, L. D. and Lifshitz, E. M. (1980). *Course of Theoretical Physics, Statistical Physics*. Butterworth-Heinemann, 5, Third Edition.

[4] Pathria, R. K. and Beale, P. D. (2011). *Statistical Mechanics*, Third Edition, Academic Press.

[5] Planck, M. (1900). *Annalen der Physik*. **2**:202-204. **2**:237(1900).

[6] Planck, M. (1900). *Annalen der Physik*. **306**(4), 719-737.

[7] Planck, M. (1900). *Annalen der Physik*. **306**(1), 69-122.

[8] Planck, M. (1901). *Annalen der Physik*, **309**(3), 553-563.

[9] Einstein, A. (1905). *Annalen der Physik*, **322**(6), 132-148.

[10] Debye, P. (1912). *Annalen der Physik*, **39**(4), 789-839.

[11] Einstein, A. (1907). *Annalen der Physik*, **22**, 180-190.

[12] Balmer, J. J. (1885). *Annalen der Physik*, **261**(5), 80-87.

[13] Gerlach, W. and Stern, O. (1922). *Zeitschrift für Physik*. **9**, 349-352, Zeitschrift für Physik. **9**, 353-355, Zeitschrift für Physik. **8**, 110-111.

[14] Compton, A. H. (1923). *Phys. Rev.* **21**(5), 483-502.

[15] Davidson, C. J. and Germer, L. H. (1928). *PNAS*. **14**(4), 317-322.

[16] Davidson, C. and Germar, L. H. (1927). *Nature*. **119**(2998), 558.

[17] Thomson, G. P. (1928). *Proceedings of the Royal Society of London*. Series A, **117**(778), 600-609.

[18] Raman, C. V. (1928). *Ind. J. Phys.*, **2**, 387-398.

[19] Raman, C. V. and Krishnan, K. S. (1928). *Nature* **121** (3048), 501-502.

[20] Bohr, N. (1913). *Philosophical Magazine*. **26**(151), 1-24.

[21] Bohr, N. (1913). *Philosophical Magazine*. **26**(153), 476-502.

[22] Bohr, N. (1913). *Philosophical Magazine*. 857-875.

[23] Bohr, N. (1921). *Nature*. **92**(2295), 232-232 (1914), **107**(2682), 104-107.

[24] Wilson, W. (1915). *Philosophical Magazine*. **29**(174), 795-802.

[25] Sommerfeld, A. (1916). *Annalen der Physik*. **51**(17), 1.

[26] Schrodinger, W. (1925). *Zeitschrift für Physik*. **33**, 879-893.

[27] Schrödinger, E. (1926). *Phys. Rev.* **28(6)**, 1049-1070.

[28] Dirac, P. A. M. (1928). *Proceedings of the Royal Society A: Physical and Engineering Sciences*. **117(778)**, 610.

[29] Dirac, P. A. M. (1930). *Proceedings of the Royal Society A: Physical and Engineering Sciences*, **126(801)**, 360.

[30] Heisenberg, W. (1930). *Zeitschrift für Physik.* **43** (4-3), 172-198.

[31] de Bröglie, L. (1925). *Ann. Phys.* (Paris), **3 (22)**.

[32] Pilar, F. L. (1990). *Elementary Quantum Chemistry.* Second Edition, McGraw-Hill.

[33] Levine, I. N. (2000). *Quantum Chemistr*y. Fifth Edition, Pearson Education Inc.

[34] Merzbecher, E. (1970). *Quantum Mechanics*. Second Edition, Wiley.

[35] Messiah, A. (1963).*Quantum Mechanics*. Vols. 1 and 2, Halsted.

[36] Atkins, P. W. and Friedman, R. S. (1997). *Molecular Quantum Mechanics*. Third Edition, Oxford University Press.

[37] Bohr, N. (1920). *Zeitschrift für Physik*, **2(5)**, 423-478.

[38] Ehrenfest, P. (1927). *Z. Physik*, **45**, 455-457.

[39] Ruark, A. E. (1928). *Phys. Rev.* **31**, 533.

[40] Jammer, M. (1966). *The Conceptual Development of Quantum Mechanics*. McGraw-Hill.

[41] Wheeler, N. (1998). *Remarks concerning the status and some ramifications of Ehrenfest Theorem. Lecture notes*, 1-37.

[42] Ritz, W. (1908). *Annalen der Physik*, **330**(4), 660-696.

[43] Fock, V. (1930). *Zeitschrift für Physik A*, **63**(11), 855-858.

[44] Bell, R. P. (1980). *The Tunnel Effect in Chemistry*. First Edition, Chapman and Hall.

[45] Clausius, R. J. E. (1870). *Philosophical Magazine*. Series A, **40**, 122-127.

[46] Güttinger, P. (1932). *Zeitschrift für Physik*, **73**(3-4), 169.

[47] Pauli, W. (1933) Die allgemeine Prinzipien der Wellenmechanick [The general principles of wave mechanics]. In: Geiger, H. and Scheel, K., Eds., *Handbuch der Physik*, Vol. 24, Part 1, Springer, Berlin, p: 162.

[48] Feynman, R. P. (1939). *Physical Review*, **56**(4), 340.

[49] Schrödinger, E. (1926). "Der stetige bergang von der Mikrczur Makromechanik, ["The steady transition from micro to macro mechanics"] *Die Naturwissenschaften*, Vol. 14, **28**, 664-666.

In: Harmonic Oscillators
Editor: Yilun Shang
ISBN: 978-1-53615-810-6

Chapter 3

LATTICE DYNAMICS IN CRYSTALS: VIBRATIONS OF ATOMS AS SIMPLE HARMONIC OSCILLATORS

Neeraj Mehta*
Physics Department, Institute of Science,
Banaras Hindu University, Varanasi, India

ABSTRACT

In early 1900, the great German physicist Max Planck treated the atoms of a black body as simple harmonic oscillators (SHO) when he proposed the theory of blackbody radiation as a pioneering result of modern physics and quantum theory. This theory put forward an idea that the dynamics of atoms in a crystal are important. Later this idea was proved helpful in the explanation of the failure of the static lattice approximation.In this chapter; we discuss the nature of atomic motions, sometimes referred to as lattice vibrations. The lattice dynamics deals with the vibrations of the atoms inside the crystals. In order to write the dynamic equations of the motion of crystal atoms, we need to describe an inter-atomic interaction. Therefore, it is natural to start the study of the lattice dynamics with the case of small harmonic vibrations.

The dynamics of one-dimensional and two-dimensional vibrations of monatomic and diatomic crystals can be understood by using the simple model

* Author's E-mail: dr_neeraj_mehta@yahco.co.in.

forces based on harmonic approximation. This harmonic approximation is related to a simple ball-spring model. According to this model, each atom is coupled with the neighboring atoms by identical springs of same force constants. The collective motion of atoms leads to a distinct traveling wave over the whole crystal, leading to the collective motion, so-called phonon. The simple ball-spring model enlightens us some of the significant common features of lattice dynamics that have been discussed throughout this chapter. Further, this chapter helps in understanding the quantization energy of a harmonic oscillation and the concept of phonon. This chapter is the results of seven year teaching carried out by the author in the second semester of MSc program of his department.

The chapter starts with the brief introduction in section 1 to explain the reasons behind the failure of static lattice model and the concept of the ball-spring model of crystal atoms. The sub-sections deal with some initial fundamental problems of linear harmonic vibrations of a finite number of atoms as coupled oscillators. Section 2 describes the need of the harmonic approximation. Further, we discuss the lattice vibrations of the one-dimensional mono-atomic chain of infinite length and finite length in the sub-sections 2.1 and 2.2 respectively. The dispersion relation of such system of atomic harmonic oscillators with nearest neighbor interaction and the long-range interactions has been established in section 3. The relation has been also determined for the corresponding force constants.

Sections 4 and 5 deal with the lattice vibrations of the one-dimensional diatomic chain where there are two atoms per primitive cell. Particularly, we discuss two cases in these sections. Section 4 is based on the first case where the two atoms in the primitive cell are not identical i.e., their masses are not the same but they are coupled with the springs of same force constant. Section 5 is based on the second case where the two atoms in the primitive cell are identical and therefore the masses of both atoms are same, but they are coupled with the springs of unlike force constant. The acoustical and optical modes of atomic vibrations have been also illustrated in this section.

The lattice vibrations of simple two-dimensional and three-dimensional atomic chains have been discussed in sections 6 and 7. Section 8 reveals that the atomic vibrations in a crystal can be quantized and this quantization gives rise to the concept of phonon.

1. INTRODUCTION

In solid state physics, many physical phenomena can be understood and the corresponding physical properties can be determined on the basis of the static lattice model. In this model, the atoms of a crystalline solid are arranged in a fixed, rigid and immobile periodic array. In the classical theory, the static lattice model can be valid only at absolute zero temperature at which any kind of motion of all atoms ceases. At a non-zero temperature, each atom or ion must have some thermal energy and therefore a certain amount of motion in the vicinity of its rest position. However, the static lattice model is incorrect in the quantum theory even at absolute zero temperature due to the existence of zero-point energy.

1.1. Failure of the Static Lattice Model

In general, we can separate the physical properties of a crystalline solid in two categories. In the first category, we consider those physical properties that are dependent on electrons. On the other hand, the second category covers such physical properties that are related to the movement of the crystal atoms about their rest positions.

Thus, there are a large number of physical phenomena that cannot be understood with the help of the static lattice model. The examples of the failure of the static lattice model are pointed below:

- Effect of temperature on the crystal lattice, e.g., thermal expansion
- The existence of the phase transition like melting
- Transport properties like thermal conductivity and sound propagation
- The dielectric phenomenon at low frequencies
- Superconductivity and its consequences
- Interaction of radiation with the matter

1.2. Dynamic Lattice Model

At a non-zero temperature, the movements of the valence electrons are much faster than that of the corresponding atomic nuclei due to low masses of electrons as compared to nuclei. Thus, the categorization of physical properties into the properties associated with atomic vibrations and electronic properties is qualitatively straightforward to give the explanation for different physical phenomena. This concept forms the basis of the dynamic lattice model.

According to the dynamic lattice model, any atom at a lattice point of a crystal lattice is connected to its neighboring atoms by the elastic springs. Thus, any excitation produced in an atom by an external force is shared by all the atoms and all the atoms become simple harmonic oscillators. Thus, the whole crystal lattice vibrates under the influence of the external forces.

1.3. Ball-Spring Model of a Harmonic Crystal

When the atoms of a crystal are displaced from their equilibrium positions (i.e., rest positions) under the influence of thermal agitations or any other external perturbation, then each atom experiences a net force in the form of a restoring force. According to ball-spring model, we can treat the entire crystal as a harmonic system consisting of a large number of balls. Each ball is connected with its neighboring ball with the help of elastic springs as shown in Figure 1. In this model, the ball represents the atoms, while the springs represent the coupling bonds (i.e., interaction forces) between them.

If a disturbance is created in any atom under the influence of an external force then it is shared by all atoms because of linking of each atom with its neighboring atoms by elastic springs. Thus, all the atoms in the crystal behave like simple harmonic oscillators and the crystal is known as a harmonic crystal.

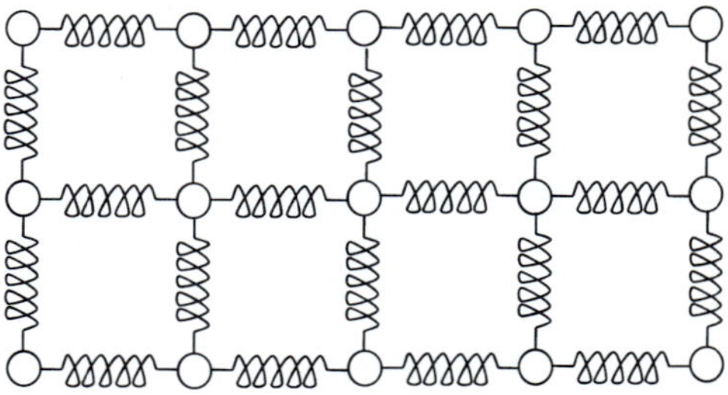

Figure 1. Illustration of ball-spring model.

Let us consider some examples of atom-spring assemblies in next sub-section to understand the complexities of mathematical formation of atom-atom interaction when more and more atoms are connected with each other by springs. Some general examples of the ball-spring model are discussed in sub-sections.

1.3.1. Longitudinal Oscillations of One Atom Coupled with Two Identical Springs

Consider the longitudinal oscillations of one atom coupled with two identical springs having the same spring constant C as shown in Figure 2.

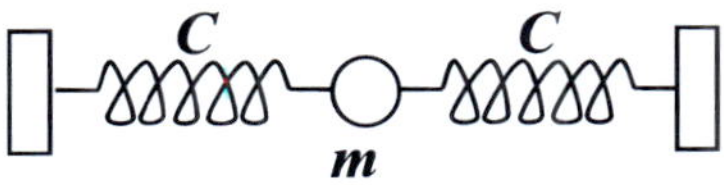

Figure 2. Longitudinal oscillations of one atom coupled with two identical springs

The equation of motion for the longitudinal oscillations of this assembly can be expressed as:

$$m.\frac{d^2x}{dt^2} = -Cx - Cx = -2Cx \tag{1}$$

Here m is the mass of the atom.

Suppose that its general solution is:

$$x = Ae^{i\omega t} \tag{2}$$

Here A is constant.

From equations (1) and (2), we obtain the general differential equation of SHM:

$$\frac{d^2x}{dt^2} + \omega^2 x = 0$$

Here $\omega = \sqrt{\frac{C}{m}}$ is the frequency of the longitudinal oscillations of the present system.

1.3.2. Longitudinal Oscillations of Two Atoms Coupled with Three Identical Springs

Consider the longitudinal oscillations of two atoms coupled with three identical springs having the same spring constant *C* as shown in Figure 3.

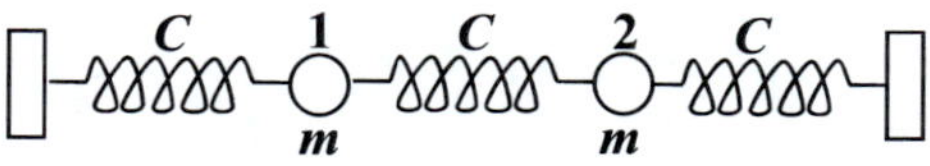

Figure 3. Longitudinal oscillations of two atoms coupled with three identical springs.

Suppose that x_1 and x_2 are the instantaneous position of atoms 1 and 2 at any instant *t* during SHM. The equations of motion for the longitudinal oscillations of this assembly can be expressed as:

$$m.\frac{d^2x_1}{dt^2} = -Cx_1 - Cx_1 + Cx_2 = -Cx_1 + C(x_2 - x_1) \tag{3i}$$

$$m.\frac{d^2x_2}{dt^2} = -Cx_2 - Cx_2 + Cx_1 = -Cx_2 - C(x_2 - x_1) \tag{3ii}$$

Here m is the mass of each atom.

Suppose that the general solutions of these equations are:

$$x_1 = A_1 e^{i\omega t} \tag{4i}$$

$$x_2 = A_2 e^{i\omega t} \tag{4ii}$$

Here A_1 and A_2 are constants.

From equations (3) and (4), we obtain the general differential equations of SHM:

$$\frac{d^2x_1}{dt^2} + \omega^2 x_1 = 0$$

$$\frac{d^2x_2}{dt^2} + \omega^2 x_2 = 0$$

Here ω is the frequency of the longitudinal oscillations of present system. To determine the values of ω, we substitute the values of x_1 and x_2 from equation (4) to (3) so that:

$$-m\omega^2 x_1 = -2Cx_1 + Cx_2 \text{ i.e.,} (2C - m\omega^2)x_1 = Cx_2 \tag{5i}$$

and

$$-m\omega^2 x_2 = -2Cx_2 + Cx_1 \text{ i.e.,} (2C - m\omega^2)x_2 = Cx_1 \tag{5ii}$$

Solving (5*i*) and (5*ii*), we get:

$$\frac{x_1}{x_2} = \frac{C}{(2C - m\omega^2)} = \frac{(2C - m\omega^2)}{C} \tag{6}$$

or

$$(2C - m\omega^2)^2 = C^2$$

or

$$2C - m\omega^2 = \pm C$$

or

$$\omega = \sqrt{\frac{C}{m}}, \sqrt{\frac{3C}{m}}$$

Thus, there are two modes of vibrations.

Substituting these values of ω in (6), the correlations between x_1 and x_2 for the two modes of vibrations are as follows:

$$x_1 = x_2 \text{and } x_1 = -x_2$$

1.3.3. Longitudinal Oscillations of Three Atoms Coupled with Four Identical Springs

Consider the longitudinal oscillations of three atoms coupled with four identical springs having the same spring constant *C* as shown in Figure 4.

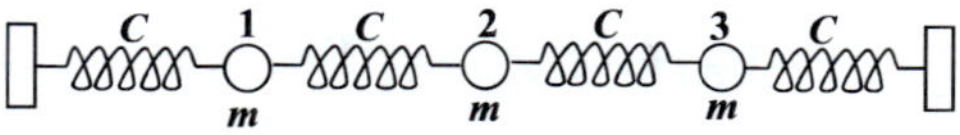

Figure 4. Longitudinal oscillations of three atoms coupled with four identical springs.

Suppose that x_1, x_2 and x_3 are the instantaneous position of atoms 1, 2 and 3 at any instant *t* during SHM. The equations of motion for the longitudinal oscillations of this assembly can be expressed as:

$$m.\frac{d^2x_1}{dt^2} = -Cx_1 - Cx_1 + Cx_2 = -2Cx_1 + Cx_2 \quad (7i)$$

$$m.\frac{d^2x_2}{dt^2} = -Cx_2 - Cx_2 + Cx_1 + Cx_3 = -2Cx_2 + Cx_1 + Cx_3 \quad (7ii)$$

$$m.\frac{d^2x_3}{dt^2} = -Cx_3 - Cx_3 + Cx_2 = -2Cx_3 + Cx_2 \quad (7iii)$$

Here m is the mass of each atom.

Suppose that the general solutions of these equations are:

$$x_1 = A_1e^{i\omega t} \quad (8i)$$

$$x_2 = A_2e^{i\omega t} \quad (8ii)$$

$$x_3 = A_3e^{i\omega t} \quad (8iii)$$

Here A_1, A_2 and A_3 are constants.

From equations (7) and (8), we obtain the general differential equations of SHM:

$$\frac{d^2x_1}{dt^2} + \omega^2 x_1 = 0$$

$$\frac{d^2x_2}{dt^2} + \omega^2 x_2 = 0$$

$$\frac{d^2x_3}{dt^2} + \omega^2 x_3 = 0$$

Here ω is the frequency of the longitudinal oscillations of present system. To determine the values of ω, we substitute the values of x_1, x_2 and $x_3$from equation (8) to (7). Then the equations (7*i*), (7*ii*) and (7*iii*) take the following forms:

$$(2C - m\omega^2).x_1 - C.x_2 + 0.x_3 = 0 \tag{9i}$$

$$C.x_1 + (m\omega^2 - 2C).x_2 + Cx_3 = 0 \tag{9ii}$$

$$0.x_1 + C.x_2 + (m\omega^2 - 2C)x_3 = 0 \tag{9iii}$$

Solving (9*i*), (9*ii*) and (9*iii*) using the determinant method, we have

$$\begin{vmatrix} (2C - m\omega^2) & -C & 0 \\ C & (m\omega^2 - 2C) & C \\ 0 & C & (m\omega^2 - 2C) \end{vmatrix} = 0$$

i.e., $(2C - m\omega^2)\{(m\omega^2 - 2C)^2 - C^2\} + C\{C(m\omega^2 - 2C) - 0\} = 0$
i.e., $(2C - m\omega^2)\{(m\omega^2 - 2C)^2 - C^2\} - C^2(2C - m\omega^2) = 0$
i.e., $(2C - m\omega^2)[(m\omega^2 - 2C)^2 - 2C^2] = 0$
i.e., either $(2C - m\omega^2) = 0$ or $[(m\omega^2 - 2C)^2 - 2C^2] = 0$

Thus, the possible values of frequencies are:

$$\omega = \sqrt{\frac{2C}{m}}, \sqrt{2 \pm \sqrt{2}} \; and \sqrt{\frac{C}{m}}$$

Thus, there are three modes of vibrations.

Substituting these values of ω in (9), the correlations between x_1, x_2 and $x_3$for the three modes of vibrations are as follows:

$$x_1 : x_2 : x_3 = \begin{Bmatrix} 1:0:-1 \\ 1:-\sqrt{2}:1 \\ 1:\sqrt{2}:1 \end{Bmatrix}$$

2. Harmonic Approximation

From above examples discussed in the sub-sections of section 1.3, one can easily realize the complexity arising in the analysis of the lattice vibrations of atoms in a crystal when we want to study the problem quantitatively by its mathematical formulation. Our calculations will be restricted to the lattice vibrations of small amplitude. Since the crystal lattice is then near to a position of steady equilibrium, therefore its movement can be determined by a generalization of the method used to analyze a simple harmonic oscillator. The small amplitude limit is known as the harmonic limit or harmonic approximation.

In the linear region (i.e., the region of elastic deformation), the restoring force on each atom is approximately proportional to its displacement (Hooke's Law). There are some effects of non-linearity or 'an harmonicity' for larger atomic displacements.

2.1. Lattice Vibrations of a One-Dimensional Mono-Atomic Chain of Infinite Length

The simplest problem of lattice dynamics of a crystal can be considered by assuming the one-dimensional chain of identical atoms. Such chain consists of a very large number of identical atoms with same masses. All the atoms in the chain are separated by the same distance with its neighboring atoms. This distance is known as equilibrium distance or lattice constant. Atoms move only in a direction parallel to the chain. For convenience, first, we consider only the interaction of each atom with its nearest neighboring atoms (short-range forces).

Let us consider a one-dimensional mono-atomic chain of the identical atoms of mass M and the infinite length of equilibrium distance or lattice constant a. Suppose that this chain is situated along the X-axis. According to atom-spring model, we can consider the interaction between two consecutive atoms in the form of an elastic spring of force constant C. Suppose that x_n and x_{n+m} are the rest positions of nth and (n+m)th atoms from the origin O at time t = 0 (i.e., when the crystal lattice is stationary) as shown in Figure 5.

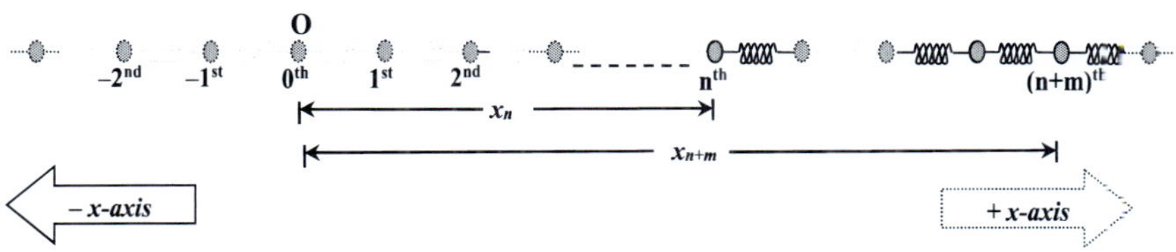

Figure 5. Lattice vibrations of one-dimensional mono-atomic chain of infinite length.

From Figure 5, it is clear that the values of rest positions of nth and (n+m)th atoms can be expressed as:

$$x_n = na \tag{10i}$$

and

$$\boldsymbol{x_{n+m} = (n+m)a} \tag{10ii}$$

If u_n and u_{n+m} are the instantaneous displacements of nth and (n+m)th atoms respectively at any instant "t" during the lattice vibrations, then their corresponding position vectors can be written as:

$$\vec{r}_n = (x_n + u_n)\hat{\imath} = (na + u_n)\hat{\imath} \tag{11i}$$

and

$$\vec{r}_{n+m} = (x_{n+m} + u_{n+m})\hat{\imath} = \{(n+m)a + u_{n+m}\}\hat{\imath} \tag{11ii}$$

Subtracting (11*i*) from (11*ii*), the instantaneous relative position for nth and (n+m)th atoms is:

$$|\vec{r}_{n+m} - \vec{r}_n| = ma + (u_{n+m} - u_n) \tag{12}$$

The interaction energy $U_{n,n+m}$ between the nth and (n+m)th atoms at any instant depends upon their instantaneous relative position, i.e.,$U_{n,n+m} = U(|\vec{r}_{n+m} - \vec{r}_n|)$.

Therefore the potential energy of the whole system may be expressed as:

$$V = \sum_{n,m} U_{n,n+m} = \sum_{n,m} U(|\vec{r}_{n+m} - \vec{r}_n|) = \sum_{n,m} U\{ma + (u_{n+m} - u_n)\}$$

Taking the Taylor expansion of R.H.S. of above equation about the point *ma* keeping in mind that $u_n \ll a$, we have

$$V = \sum_{n,m} U(ma) + \sum_{n,m} U'(ma).(u_{n+m} - u_n) + \sum_{n,m} \frac{U''(ma)}{2!}.(u_{n+m} - u_n)^2 + \sum_{n,m} \frac{U'''(ma)}{3!}.(u_{n+m} - u_n)^3 + \_\_\_\_\_\_ \tag{13}$$

But $U(ma) = \text{constant}$, therefore $U'(ma) = 0$.

Thus, we can write

$$\sum_{n,m} U(ma) = \text{constant} = V_0(\text{say})$$

Here V_0 represents the potential energy of the crystal lattice corresponding to equilibrium configuration.

If we neglect the terms containing $(u_{n+m} - u_n)^3$and higher powers of $(u_{n+m} - u_n)$, then this approximation is known as harmonic approximation as discussed in section 2.0. Now the equation (13) reduces to the following form:

$$V = V_0 + \sum_{n,m} \frac{U''(ma)}{2}.(u_{n+m} - u_n)^2 \tag{14}$$

Further let us consider the interaction of each atom with its nearest neighboring atoms only. Then, the subscript "m" drops and the equation (14) takes the form:

$$V = V_0 + \frac{U''(a)}{2}\sum_n (u_{n+1} - u_n)^2 \tag{15}$$

Now, if we consider the vibrations of pth atom, then its equation of motion can be expressed as:

$$M\frac{\partial^2 u_p}{\partial t^2} = -\frac{\partial V}{\partial u_p} \tag{16}$$

Using equation (16) in (15), we have

$$M\frac{\partial^2 u_p}{\partial t^2} = -\frac{\partial}{\partial u_p}\left\{V_0 + \frac{U''(a)}{2}\sum_n (u_{n+1} - u_n)^2\right\},$$

$$M\frac{\partial^2 u_p}{\partial t^2} = -\frac{U''(a)}{2}\frac{\partial}{\partial u_p}\{\sum_n (u_{n+1} - u_n)^2\},$$

$$M\frac{\partial^2 u_p}{\partial t^2} = -\frac{U''(a)}{2}\frac{\partial}{\partial u_p}\left\{\left(u_p - u_{p-1}\right)^2 + \left(u_{p+1} - u_p\right)^2\right\},$$

$$M\frac{\partial^2 u_p}{\partial t^2} = -\frac{U''(a)}{2}\{2.\left(u_p - u_{p-1}\right) + (-2).\left(u_{p+1} - u_p\right)\},$$

$$M\frac{\partial^2 u_p}{\partial t^2} = U''(a)\{u_{p+1} + u_{p-1} - 2u_p\} \tag{17}$$

Suppose that the general solution of equation (17) is:

$$u_p = A.e^{-i(\omega t - kx_p)} = A.\exp\{-i(\omega t - kpa)\} \tag{18}$$

Here $x_p = pa$ is the rest position of pth atom from the origin O and $k = 2\pi / \lambda$ is the propagation constant of the lattice wave of wavelength λ.

Then we can easily see that:

$$\frac{\partial u_p}{\partial t} = -i\omega A.e^{-i(\omega t - kx_p)} = -i\omega u_p$$

So that:

$$\frac{\partial^2 u_p}{\partial t^2} = -i\omega \frac{\partial u_p}{\partial t} = -i\omega.\left(-i\omega u_p\right) = -\omega^2 u_p$$

Also, we have

$$u_{p+1} = A.\exp[-i\{\omega t - kx_{p+1}\}]$$

or $u_{p+1} = A.\exp[-i\{\omega t - k(p+1)a\}]$
or $u_{p+1} = A.\exp[-i\{\omega t - kpa\}].e^{ika}$
or $u_{p+1} = e^{ika}u_p$

Similarly

$$u_{p-1} = e^{-ika}u_p$$

Thus, the equation (17) becomes:

$$M(-\omega^2 u_p) = U''(a)\left\{e^{ika}u_p + e^{-ika}u_p - 2u_p\right\}$$

or $-M\omega^2 u_p = U''(a)\left\{e^{ika} + e^{-ika} - 2\right\}u_p$
or $-M\omega^2 = U''(a)\{e^{ika/2} - e^{-ika/2}\}^2$
or $-M\omega^2 = U''(a)\{\, 2i.sin\left(\frac{ka}{2}\right)\}^2$
or $-M\omega^2 = -4U''(a)sin^2\left(\frac{ka}{2}\right)$

Thus, the angular frequency for the lattice vibrations of one-dimensional mono-atomic chain of infinite length can be expressed as:

$$\omega = \pm\sqrt{\frac{4U''(a)}{M}}\, sin\left(\frac{ka}{2}\right)$$

Since ω is essentially a positive quantity, irrespective of whether k is positive or negative, so we consider only positive value of ω, i.e.,

$$\omega = \sqrt{\frac{4U''(a)}{M}}\, sin\left(\frac{ka}{2}\right)$$

or

$$\omega = 2\sqrt{\frac{C}{M}}\, sin\left(\frac{ka}{2}\right) \quad (19)$$

Here $U''(a) = C$ represents the spring constant. Equation (19) shows the relation between ω and k and it is known as dispersion relation. The corresponding plot of ω against k is known as dispersion curve (see Figure 6). The region $-\pi/a<k<\pi/a$ is known as the first Brillouin zone.

The dispersion relation also shows that there exists a following maximum frequency, ν_{max}, with which the lattice wave can be propagated through the chain:

$$\nu_{max} = \frac{\omega_{max}}{2\pi} = \frac{1}{\pi}\sqrt{\frac{C}{M}}$$

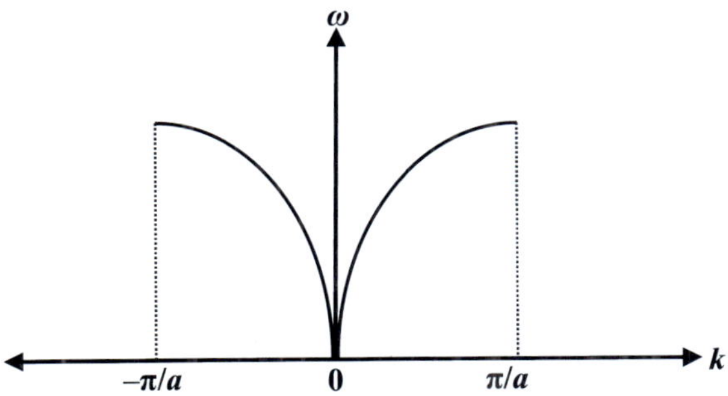

Figure 6. Dispersion curve for the vibrations of one-dimensional mono-atomic chain.

Thus, the mono-atomic lattice acts as a low pass filter which allows the transmission of all the frequencies in the range $0 < \nu < \nu_{max}$ but blocks other frequencies.

Now we will discuss the consequences of dispersion relation in next sub-sections.

2.1.1. Dispersion Relation at Low Frequencies

At low frequencies (or high wavelengths), $k \to 0$, so we have:

$$sin\left(\frac{ka}{2}\right) \approx \frac{ka}{2}$$

Then the equation (19) reduces to:

$$\omega = 2\sqrt{\frac{C}{M}}\frac{ka}{2} = ka\sqrt{\frac{C}{M}} \tag{20}$$

The phase velocity v_p and the group velocity v_g are, therefore, given by:

$$v_p = \frac{\omega}{k} = a\sqrt{\frac{C}{M}} = \text{constant} \tag{21i}$$

$$v_g = \frac{d\omega}{dk} = a\sqrt{\frac{C}{M}} = \text{constant} \tag{21ii}$$

Therefore, the phase velocity v_p and the group velocity v_g are same at low frequencies (or high wavelengths). Thus, the lattice behaves like a continuous and homogeneous medium (i.e., non-dispersive medium).

2.1.2. Dispersion Relation at High Frequencies

At high frequencies (or low wavelengths), the phase velocity v_p and the group velocity v_g can be expressed as:

$$v_p = \frac{\omega}{k} = \frac{2}{k}\sqrt{\frac{C}{M}}\,sin\left(\frac{ka}{2}\right) = a\sqrt{\frac{C}{m}}\frac{sin\left(\frac{ka}{2}\right)}{\frac{ka}{2}} \tag{22i}$$

$$v_g = \frac{d\omega}{dk} = a\sqrt{\frac{C}{M}}\,cos\left(\frac{ka}{2}\right) \tag{22ii}$$

Therefore, the phase velocity v_p and the group velocity v_g are not the same at high frequencies (or low wavelengths). Thus, the lattice behaves like a dispersive medium. The plots of the phase velocity v_p and the group velocity v_g against k are shown in Figures 7 and 8 respectively.

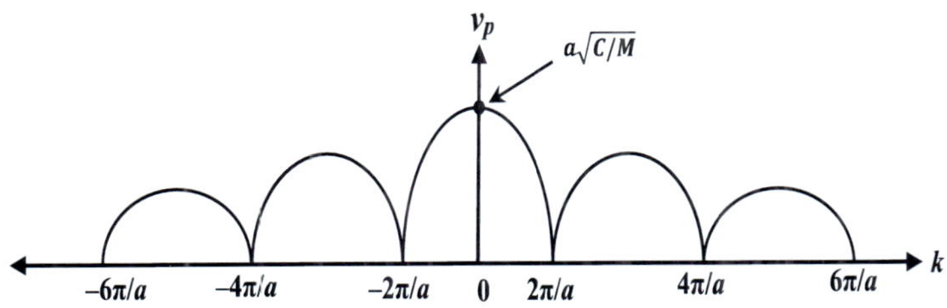

Figure 7. Plot of phase velocity v_p against k.

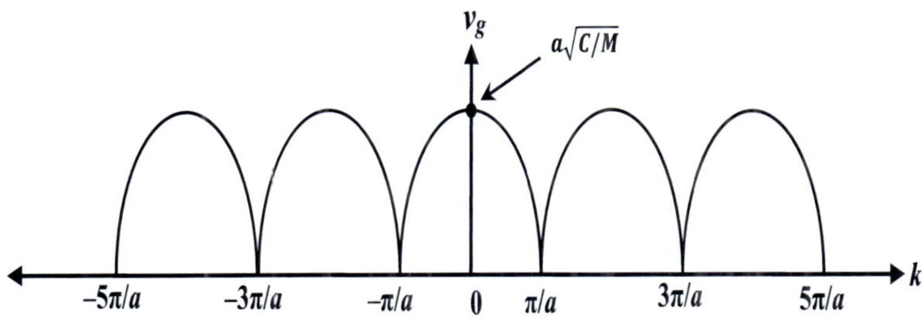

Figure 8. Plot of group velocity v_g against k.

2.1.3. Condition for the Standing Wave: Equivalence with Bragg's Reflection

From equation (19), it is clear that the maximum value of ω is $2(C/M)^{1/2}$ which is obtained for the condition:

$$sin\left(\frac{ka}{2}\right) = 1$$

i.e.,
$$sin\left(\frac{ka}{2}\right) = sin\frac{(2l+1)\pi}{2}$$

i.e.,
$$\frac{ka}{2} = \frac{(2l+1)\pi}{2}$$

i.e., $$k = \frac{(2l+1)\pi}{a}; l = 0,1,2,3,4 \ldots \ldots \ldots \ldots$$

The minimum value of k is given by:

$$k = \frac{\pi}{a}$$

or

$$\frac{2\pi}{\lambda} = \frac{\pi}{a} \text{ [since } k = 2\pi / \lambda]$$

or

$$\lambda = 2a \tag{23}$$

For the maximum value of ω is $2(C/M)^{1/2}$, the phase velocity v_p and the group velocity v_g are:

$$v_p = \frac{\omega}{k} = \text{finite} \tag{24i}$$

$$v_g = \frac{d\omega}{dk} = 0 \tag{24ii}$$

Equation (22*ii*) reveals that the group velocity v_g will be zero for the following condition:

$$cos\left(\frac{ka}{2}\right) = 0$$

i.e., $$cos\left(\frac{ka}{2}\right) = cos\,\frac{(2l+1)\pi}{2}$$

i.e., $$\frac{ka}{2} = \frac{(2l+1)\pi}{2}$$

i.e., $$k = \frac{(2l+1)\pi}{a}; l = 0,1,2,3,4 \ldots \ldots \ldots \ldots$$

Since the group velocity v_g represents the velocity with which the energy is transferred, therefore the equation (24*ii*) indicates that no energy is transferred at the frequency $2(C/M)^{1/2}$. Thus, we have a standing wave. This condition is similar to the Bragg's condition for the reflection of X-rays in a crystal. According to Bragg's condition, there is no propagation of a traveling wave inside the crystal due to the reflection of X-rays from the consecutive atomic planes. If d is the distance between the consecutive atomic planes and λ is the wavelength of X-rays incident at the glancing angle θ, then Bragg's condition can be expressed as:

$$n\lambda = 2d\, sin\theta$$

For first order reflection (i.e., n = 1) and $\theta = 90^{o}$, we have

$$\lambda = 2d \tag{25}$$

Equation (23) is equivalent to the equation (25) i.e., Bragg's condition for the reflection. Thus, at $k = \pi/a$, the alternate atoms oscillate in opposite phases as shown in Figure 9. Consequently, a standing wave is formed as a signature of zero group velocity v_g.

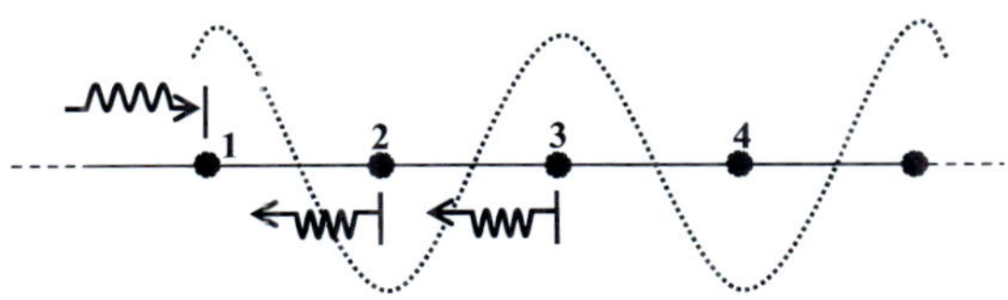

Figure 9. Scattering of the wavelets from the neighboring atoms.

In Figure 9, we see that the wavelets scattered from the neighboring atoms are out of phase by an amount π. But when the wavelet reflected from atom 3reaches that reflected from 2, then these two wavelets are in phase. Similarly, this condition is applicable to other wavelets at $k = \pi/ a$. Consequently, all the scattered wavelets interfere constructively, and thus the maximum reflection takes place. Thus, this situation applied to elastic waves is equivalent to the Bragg condition. Thus, it is clear that the wave cannot propagate in the lattice, but through the successive reflections back and forth, a standing wave is set up.

2.3. Lattice Vibrations of a One-Dimensional Mono-Atomic Chain of Finite Length

Now let us consider a one-dimensional mono-atomic chain of a finite length of equilibrium distance or lattice constant *a* situated along the X-axis. Suppose that there are N+1 atoms in the chain and they are leveled as 0th, 1st, 2nd, 3rd, ... Nth atoms from the origin O as shown in Figure 10. Then the length of this chain will be *aN*. Suppose that the 0th end Nth atoms located at the origin and at a distance *aN* from the origin respectively are fixed i.e., they remain stationary during the lattice vibrations.

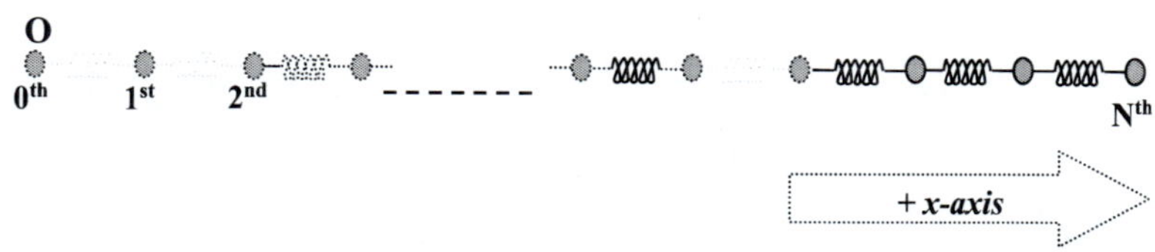

Figure 10. Lattice vibrations of one-dimensional mono-atomic chain of finite length.

If we apply the harmonic approximation and consider the nearest neighbor interaction only as we have discussed in section 2.1, then the potential energy of the whole system may be expressed as:

$$V = V_0 + \frac{C}{2}\{\sum_{n=1}^{N-2}(u_{n+1} - u_n)^2 + (u_1^2 + u_{N-1}^2)\}$$

or

$$V = V_0 + \frac{C}{2}\sum_{n=1}^{N-2}(u_{n+1} - u_n)^2 + \frac{C}{2}(u_1^2 + u_{N-1}^2) \tag{26}$$

Now, if we consider the vibrations of pth atom, then its equation of motion can be expressed as:

$$M\frac{\partial^2 u_p}{\partial t^2} = -\frac{\partial V}{\partial u_p} \tag{27}$$

Using equation (26) in (27), we have

$$M\frac{\partial^2 u_p}{\partial t^2} = -\frac{\partial}{\partial u_p}\left\{V_0 + \frac{C}{2}\sum_{n=1}^{N-2}(u_{n+1} - u_n)^2 + \frac{C}{2}(u_1^2 + u_{N-1}^2)\right\},$$

$$M\frac{\partial^2 u_p}{\partial t^2} = -\frac{C}{2}\frac{\partial}{\partial u_p}\left\{\left(u_p - u_{p-1}\right)^2 + \left(u_{p+1} - u_p\right)^2\right\},$$

$$M\frac{\partial^2 u_p}{\partial t^2} = C.\{u_{p+1} + u_{p-1} - 2u_p\} \tag{28}$$

Now since the chain is finite and bounded at both ends, so the general solution of the equation (28) can be expressed as:

$$u_p = A_1.e^{\{-i(\omega t - kx_p)\}} + A_2.e^{\{i(\omega t - kx_p)\}}$$

i.e., $u_p = A_1.\exp\{-i(\omega t - kpa)\} + A_2.\exp\{i(\omega t - kpa)\}$

Then we can easily see that:

$$\frac{\partial^2 u_p}{\partial t^2} = -\omega^2 u_p$$

Also, we have

$$u_{p+1} = A_1.\exp[-i\{\omega t - k(p+1)a\}] + A_2.\exp[i\{\omega t - k(p+1)a\}]$$

or

$$u_{p+1} = e^{-i\omega t}\{A_1.e^{ikpa}e^{ika}\} + e^{i\omega t}\{A_2.e^{-ikpa}e^{-ika}\}$$

Similarly

$$u_{p-1} = e^{-i\omega t}\{A_1.e^{ikpa}e^{-ika}\} + e^{i\omega t}\{A_2.e^{-ikpa}e^{ika}\}$$

So that

$$u_{p+1} + u_{p-1} = e^{-i\omega t} A_1 . e^{ikpa} \{ e^{ika} + e^{-ika} \} + e^{i\omega t} A_2 . e^{-ikpa} \{ e^{ika} + e^{-ika} \}$$

or $u_{p+1} + u_{p-1} = [e^{-i\omega t} A_1 . e^{ikpa} + e^{i\omega t} A_2 . e^{-ikpa}] . \{ e^{ika} + e^{-ika} \}$
or $u_{p+1} + u_{p-1} = u_p . \{ e^{ika} + e^{-ika} \}$

Thus, the equation (28) becomes:

$$-M\omega^2 u_p = C[u_p . \{ e^{ika} + e^{-ika} \} - 2u_p]$$

i.e.,$-M\omega^2 u_p = C[e^{ika} + e^{-ika} - 2]u_p$
i.e., $-M\omega^2 = C\{e^{ika/2} - e^{-ika/2}\}^2$
i.e., $-M\omega^2 = C\{ 2i. sin\left(\frac{ka}{2}\right)\}^2$
i.e., $-M\omega^2 = -4C sin^2 \left(\frac{ka}{2}\right)$
i.e., $\omega = 2\sqrt{\frac{C}{M}} sin\left(\frac{ka}{2}\right)$

Thus, again we get the same dispersion relation.

Now we will discuss an important cyclic boundary condition for the lattice vibrations of one-dimensional mono-atomic chain of finite length in next sub-section.

2.3.1. Born-von Karman's Cyclic Boundary Condition

Born and von Karman suggested a significant and useful cyclic boundary condition to represent the vibrational modes of a one-dimensional mono-atomic chain of finite length as a running wave by choosing such solutions of the problem related to one-dimensional mono-atomic chain of infinite length which are periodic with a period equal to the finite length of the chain. For this, they considered a circular chain of *N*+1 atoms such that 0th atom coincides with N^{th} atom. This condition has been illustrated in Figure 11. In such assumption, each atom is free to oscillate and the total number of the atoms is now *N*.

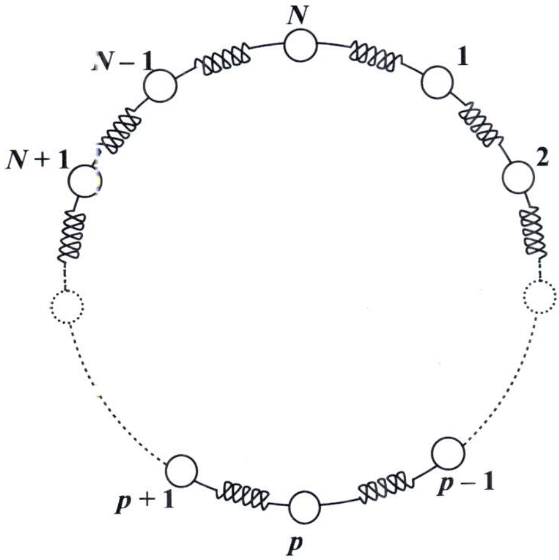

Figure 11. Illustration of Born-von Karman's cyclic boundary condition.

Thus, if If u_p and $u_{p+N}$are the instantaneous displacements of pth and (p+N)[th] atoms respectively, then 1 Born-von Karman's cyclic boundary condition can be written as:

$$u_p = u_{p+1}$$

or

$$A.\exp[-i\{\omega t - kpa\}] = A.\exp[-i\{\omega t - k(p+1)a\}]$$

or

$$\exp[ikNa] = 1$$

or

$$\exp[ikNa] = \exp[2\pi li] \ [l \text{ is an integer}]$$

or

$$k = \frac{2\pi}{Na} l; (l = 0, \pm 1, \pm 2, \pm 3, \dots \dots \dots \dots \dots)$$

Thus, the allowed values of k are as follows:

$$k = 0, \pm\frac{2\pi}{Na}, \pm 2\frac{2\pi}{Na}, \pm 3\frac{2\pi}{Na}, \pm 4\frac{2\pi}{Na}, \dots \dots \dots \dots \dots \dots$$

3. Generalized Dispersion Relation for Lattice Vibrations of the One-Dimensional Mono-Atomic Chain for the Long-Range Interactions

The generalization of the dispersion relation for the lattice vibrations of the one-dimensional mono-atomic chain can be established for the long-range interactions by considering the second, third, fourth .... nearest neighbor interactions too for each atom in addition to the first nearest neighbor interaction as shown in Figure 12.

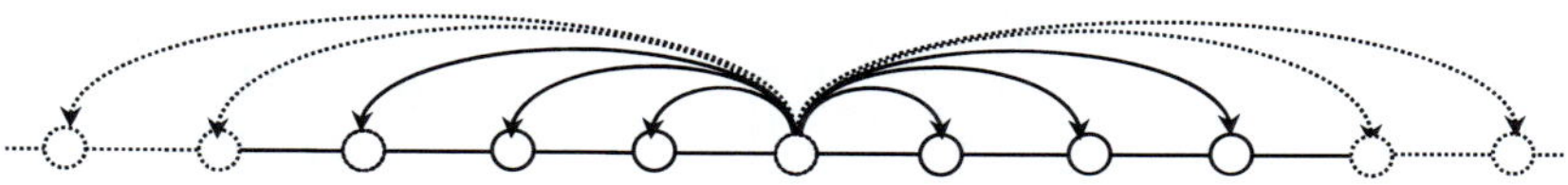

Figure 12. Interaction of a particular atom with the neighboring atoms.

Thus, if we apply the harmonic approximation but consider the interaction with all nearest neighbors for each atom, then the potential energy of the whole system may be expressed as:

$$V = V_0 + \frac{1}{2}\sum_{n,m} C_m (u_{n+m} - u_n)^2 \tag{29}$$

Now, if we consider the vibrations of pth atom, then its equation of motion can be expressed as:

$$M\frac{\partial^2 u_p}{\partial t^2} = -\frac{\partial V}{\partial u_p} \tag{30}$$

Using equation (29) in (30), we have

$$M\frac{\partial^2 u_p}{\partial t^2} = -\frac{\partial}{\partial u_p}\left\{V_0 + \frac{1}{2}\sum_{n,m} C_m (u_{n+m} - u_n)^2\right\},$$

$$M\frac{\partial^2 u_p}{\partial t^2} = -\frac{\partial}{\partial u_p}\left\{V_0 + \frac{1}{2}\sum_{m} C_m \left[\left(u_{p+m} - u_p\right)^2 + \left(u_p - u_{p-m}\right)^2\right]\right\},$$

$$M\frac{\partial^2 u_p}{\partial t^2} = \sum_m C_m.\{u_{p+m} + u_{p-m} - 2u_p\} \quad (31)$$

Consider the general solution of equation (31) as $u_p = A.e^{-i(\omega t - kx_p)}$, then we have

$$\frac{\partial^2 u_p}{\partial t^2} = -\omega^2 u_p$$

Also, we have

$$u_{p+m} = A.\exp[-i\{\omega t - kx_{p+m}\}]$$

or

$$u_{p+m} = A.\exp[-i\{\omega t - k(p+m)a\}]$$

or

$$u_{p+m} = A.\exp[-i\{\omega t - kpa\}].e^{ikma}$$

or

$$u_{p+m} = e^{ikma}u_p$$

Similarly

$$u_{p-m} = e^{-ikma}u_p$$

Thus, the equation (31) becomes:

$$M(-\omega^2 u_p) = \sum_m C_m.\{e^{ikma}u_p + e^{-ikma}u_p - 2u_p\}$$

or

$$-M\omega^2 u_p = \sum_m C_m.\{e^{ikma} + e^{-ikma} - 2\}u_p$$

or

$$-M\omega^2 = \sum_m C_m.\{e^{ikma/2} - e^{-ikma/2}\}^2$$

or

$$-M\omega^2 = \sum_m C_m.\left\{2i.sin\left(\frac{mka}{2}\right)\right\}^2$$

or

$$-M\omega^2 = -4\sum_m C_m.sin^2\left(\frac{mka}{2}\right)$$

or

$$\omega^2 = \frac{4}{M}\sum_m C_m.sin^2\left(\frac{mka}{2}\right) \quad (32)$$

or

$$\omega^2 = \frac{2}{M}\sum_m C_m.\{1-\cos(mka)\} \quad (33)$$

Equations (32) and (33) represent the generalized dispersion relation for lattice vibrations of one-dimensional mono-atomic chain for the long-range interactions.

To determine the expression for the corresponding force constants C_m, multiplying equation (33) by cos(pka) and integrating between the limits $-\pi/a$ to π/a so that:

$$\int_{-\frac{\pi}{a}}^{\frac{\pi}{a}} \omega^2.\cos(pka)dk =$$

$$\frac{2}{M}\int_{-\pi/a}^{\pi/a}\sum_m C_m.\{\cos(pka)-\cos(mka).\cos(pka)\}dk \quad (34)$$

But

$$\int_{-\pi/a}^{\pi/a}\cos(pka)dk = \left[\frac{\sin(pka)}{pa}\right]_{-\pi/a}^{\pi/a} = 0$$

Also, we have

$$\int_{-\pi/a}^{\pi/a}\cos(mka)\cos(pka)dk = \frac{1}{2}\int_{-\pi/a}^{\pi/a}2\cos(mka)\cos(pka)dk$$

$$= \frac{1}{2}\int_{-\pi/a}^{\pi/a}[\cos\{(m+p)\text{ka}\}+\cos\{(m-p)\text{ka}\}]dk$$

$$= \frac{1}{2}\left[\frac{\sin\{(m+p)\text{ka}\}}{(m+p)\text{a}} + \frac{\sin\{(m-p)\text{ka}\}}{(m-p)\text{a}}\right]_{-\pi/a}^{\pi/a} = \begin{Bmatrix} 0 \text{ for } m \neq p \\ non-zero \text{ for } m = p \end{Bmatrix}$$

For $m = p$, again we have:

$$\int_{-\frac{\pi}{a}}^{\frac{\pi}{a}} \cos(mka)\cos(pka)dk = \frac{1}{2}\int_{-\frac{\pi}{a}}^{\frac{\pi}{a}} 2\cos(mka)\cos(mka)dk$$

$$= \frac{1}{2}\int_{-\frac{\pi}{a}}^{\frac{\pi}{a}} 2\cos^2(mka)\, dk = \frac{1}{2}\int_{-\frac{\pi}{a}}^{\frac{\pi}{a}} \{\cos(2mka) + 1\}dk$$

$$= \frac{1}{2}\left[\frac{\sin(2mka)}{2ma} + k\right]_{-\pi/a}^{\pi/a} = \frac{1}{2}.\frac{2\pi}{a} = \frac{\pi}{a}$$

Therefore, the equation (34) reduces to:

$$\int_{-\pi/a}^{\pi/a} \omega^2.\cos(mka)dk = \frac{2}{M}C_m.\left\{0 - \frac{\pi}{a}\right\} = -\frac{2\pi}{Ma}C_m$$

Thus, the force constant of the one-dimensional mono-atomic chain for the long-range interactions is given by the expression:

$$C_m = -\frac{Ma}{2\pi}.\int_{-\pi/a}^{\pi/a} \omega^2.\cos(mka)dk$$

The above relation is applicable only for the one-dimensional mono-atomic chain and it determines the force constant in terms of the Fourier cosine transformation of ω^2 as a function of k.

4. Lattice Vibrations of One-Dimensional Diatomic Chain Having Alike Atoms of Different Masses Attached with Identical Springs of Same Force Constants

Let us consider the vibrations of one-dimensional diatomic lattice of lattice constant $2a$. Suppose that this one-dimensional diatomic chain has two atoms of mass m and M ($M>m$) per primitive cell. The two kinds of atoms are arranged alternately such that each atom is separated by its nearest neighbors by a distance a. Then the separations between any two successive atoms of same kind (i.e., identical atoms) will be $2a$ as shown in Figure 13.

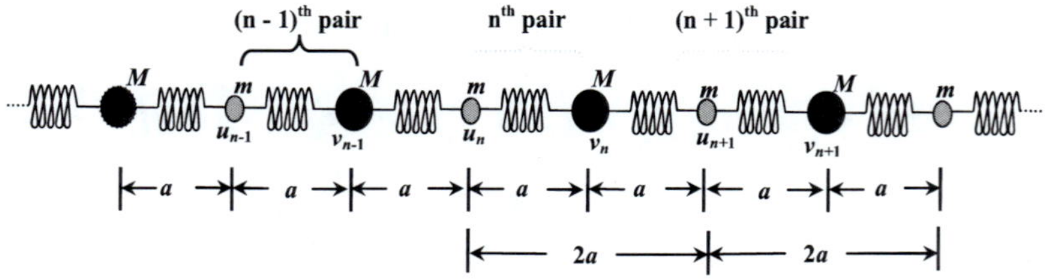

Figure 13. Lattice vibrations of one-dimensional diatomic chain.

4.1. Dispersion Relation for One-Dimensional Diatomic Chain

If we apply the harmonic approximation and consider the nearest neighbor interaction only as we have discussed in section 2.1, then the potential energy of the whole system may be expressed as:

$$V = V_0 + \frac{C}{2}\Sigma_n\{(u_{n+1} - v_n)^2 + (v_n - u_n)^2 + (u_n - v_{n-1})^2\} \quad (35)$$

Here u_n and v_n are the instantaneous displacements of atoms of mass m and M in nth pair. The equations of motions of atoms of nth pair can be, therefore, expressed as:

$$m\frac{\partial^2 u_n}{\partial t^2} = -\frac{\partial V}{\partial u_n} \quad (36i)$$

$$M\frac{\partial^2 v_n}{\partial t^2} = -\frac{\partial V}{\partial v_n} \quad (36ii)$$

In light of equations (35) and (36*i*), we have

$$m\frac{\partial^2 u_n}{\partial t^2} = -\frac{\partial}{\partial u_n}\left[V_0 + \frac{C}{2}\sum_n\{(u_{n+1} - v_n)^2 + (v_n - u_n)^2 + (u_n - v_{n-1})^2\}\right],$$

$$m\frac{\partial^2 u_n}{\partial t^2} = -\frac{C}{2}[\{2(u_n - v_{n-1}) - 2(v_n - u_n)\}]$$

$$\text{or}\, m\frac{\partial^2 u_n}{\partial t^2} = c\{v_n + v_{n-1} - 2u_n\} \tag{37i}$$

Similarly from equations (35) and (36*ii*), we have

$$M\frac{\partial^2 v_n}{\partial t^2} = c\{u_{n+1} + u_n - 2v_n\} \tag{37ii}$$

Suppose that the general solutions of equation (37) are:

$$u_n = u_0.\exp\{-i(\omega t - k.2na)\} \tag{38i}$$

and

$$v_n = v_0.\exp\{-i(\omega t - k.2na)\} \tag{38ii}$$

Here we have taken 2*na* as the rest position of nth pair atom from the origin O instead of separate distances for simplification of mathematical analysis.

Using (38*i*) in (37*i*), we have:

$$m(-\omega^2 u_n) = C\{v_n + e^{-2ika}v_n - 2u_n\}$$

or

$$(2C - m\omega^2)u_n - C(1 + e^{-2ika})v_n = 0 \tag{39i}$$

Using (38*ii*) in (37*ii*), we have:

$$M(-\omega^2 v_n) = C\{u_n + e^{2ika}u_n - 2v_n\}$$

or

$$C(1 + e^{2ika})u_n - (2C - M\omega^2)v_n = 0 \tag{39ii}$$

Solving equations (39*i*) in (39*ii*) using determinant method, we have:

$$\begin{vmatrix} (2C - m\omega^2) & -C\left(1 + e^{-2ika}\right) \\ C\left(1 + e^{2ika}\right) & -(2C - M\omega^2) \end{vmatrix} = 0$$

or

$$-(2C - m\omega^2)(2C - M\omega^2) + C^2\left(1 + e^{-2ika}\right)\left(1 + e^{2ika}\right) = 0$$

or

$$C^2\left(2 + e^{2ika} + e^{-2ika}\right) = 4C^2 + mM\omega^4 - 2C(m + M)\omega^2$$

or

$$mM\omega^4 - 2C(m + M)\omega^2 + 4C^2 = C^2\{2 + 2\cos(2ka)\}$$

or

$$mM\omega^4 - 2C(m + M)\omega^2 + 4C^2 = 2C^2\{1 + \cos(2ka)\}$$

or

$$mM\omega^4 - 2C(m + M)\omega^2 + 4C^2 = 2C^2\{2cos^2ka\}$$

or

$$mM\omega^4 - 2C(m + M)\omega^2 + 4C^2 = 4C^2cos^2ka$$

or

$$mM\omega^4 - 2C(m + M)\omega^2 + 4C^2 - 4C^2cos^2ka = 0$$

or

$$mM\omega^4 - 2C(m + M)\omega^2 + 4C^2(1 - cos^2ka) = 0$$

or

$$mM\omega^4 - 2C(m + M)\omega^2 + 4C^2sin^2ka = 0$$

Or

$$\omega^4 - \frac{2C(m+M)}{mM}\omega^2 + \frac{4C^2}{mM}sin^2ka = 0$$

$$\therefore \omega^4 - 2C\left(\frac{1}{m} + \frac{1}{M}\right)\omega^2 + \frac{4C^2}{mM}sin^2ka = 0 \qquad (40)$$

The equation (40) represents the dispersion relation correlating ω with k for the lattice vibrations of one-dimensional diatomic chain.

4.2. Dispersion Curve for the One-Dimensional Diatomic Chain

As the relation (40) is a quadratic equation of ω^2. Therefore the roots of ω^2 can be expressed as:

$$\omega^2 = \frac{2C\left(\frac{1}{m}+\frac{1}{M}\right)\pm\left[4C^2\left(\frac{1}{m}+\frac{1}{M}\right)^2-\frac{16C^2}{mM}sin^2 ka\right]^{1/2}}{2}$$

or

$$\omega^2 = C\left(\frac{1}{m}+\frac{1}{M}\right)\pm\left[C^2\left(\frac{1}{m}+\frac{1}{M}\right)^2-\frac{4C^2}{mM}sin^2 ka\right]^{1/2} \tag{41}$$

Now following two cases arise:

Case (i) When k = 0, i.e., sin ka = 0, then equation (41) reduces to:

$$\omega^2 = C\left(\frac{1}{m}+\frac{1}{M}\right)\pm C\left(\frac{1}{m}+\frac{1}{M}\right)$$

i.e.,

$$\omega^2 = \begin{cases} C\left(\frac{1}{m}+\frac{1}{M}\right)+C\left(\frac{1}{m}+\frac{1}{M}\right); \text{for positive sign} \\ C\left(\frac{1}{m}+\frac{1}{M}\right)-C\left(\frac{1}{m}+\frac{1}{M}\right); \text{ for negative sign} \end{cases}$$

i.e.,

$$\omega = \begin{cases} \sqrt{2C\left(\frac{1}{m}+\frac{1}{M}\right)} = (\omega_+)_{k=0}; \text{for positive sign} \\ 0 = (\omega_-)_{k=0}; \text{ for negative sign} \end{cases} \tag{42i}$$

Case (i) When k = π/2a, i.e., sin ka = 1, then equation (41) reduces to:

$$\omega^2 = C\left(\frac{1}{m}+\frac{1}{M}\right)\pm\left[C^2\left(\frac{1}{m}+\frac{1}{M}\right)^2-\frac{4C^2}{mM}\right]^{1/2}$$

i.e.,

$$\omega^2 = C\left(\frac{1}{m}+\frac{1}{M}\right)\pm C\left(\frac{1}{m}-\frac{1}{M}\right)$$

i.e.,

$$\omega^2 = \begin{cases} C\left(\frac{1}{m}+\frac{1}{M}\right)+C\left(\frac{1}{m}-\frac{1}{M}\right); \text{for positive sign} \\ C\left(\frac{1}{m}+\frac{1}{M}\right)-C\left(\frac{1}{m}-\frac{1}{M}\right); \text{ for negative sign} \end{cases}$$

i.e.,

$$\omega = \begin{cases} \sqrt{\frac{2C}{m}} = (\omega_+)_{k=\pi/2a}; \text{ for positive sign} \\ \sqrt{\frac{2C}{M}} = (\omega_-)_{k=\pi/2a}; \text{ for negative sign} \end{cases} \tag{42ii}$$

Thus, the corresponding dispersion curve (i.e., plot of ω against k) consists of two branches as shown in Figure 14.

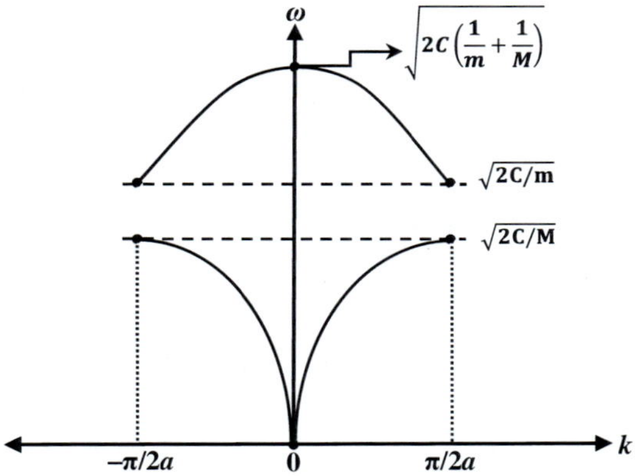

Figure 14. Dispersion curve of one-dimensional diatomic chain having atoms of alternate masses *m* and *M* connected with the identical springs of force constant *C*.

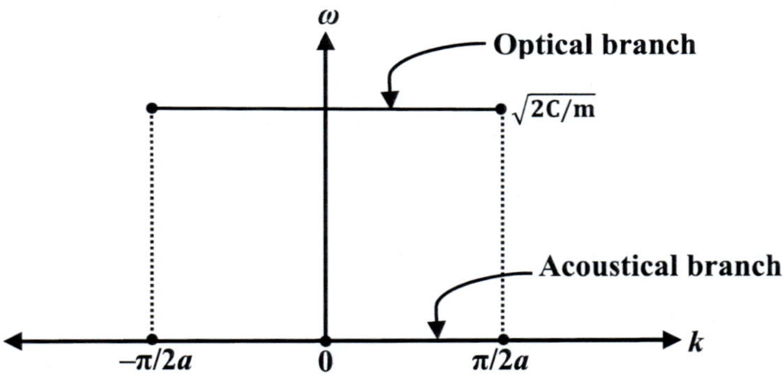

Figure 15. Dispersion curve of one-dimensional diatomic chain for M → ∞.

The allowed frequencies of propagation wave are split into an upper branch known as the optical branch, and a lower branch called the acoustical branch. The optical and acoustical branches in Figure 15 are shown by upper and the lower branches respectively. The acoustical branch looks qualitatively similar to the dispersion relation for a one-dimensional mono-atomic chain, while the optical branch shows a different kind of wave propagation.

The acoustical branch has this name because it gives rise to the long wavelength vibrations. The vibrations of acoustical modes in the one-dimensional diatomic chain take place when a sound wave passes through the crystal lattice. The optical branch consists of high energy vibrations i.e., high-frequency vibrations. The name of the optical branch comes from how these modes were discovered. We need a certain amount of energy to excite this mode. These modes can be excited by exposing the crystal lattice by the electromagnetic radiation. This refers to the situation in the ionic crystals where such a motion of the positive and the negative ions may displace the center of the positive charges with the center of the negative charges. This situation creates the electric dipoles which start the oscillations under the condition when an electromagnetic radiation is absorbed by the crystal.

4.3. Characteristics Features of the Dispersion Curve of the One-Dimensional Diatomic Chain

The characteristics features of the dispersion curve are pointed below:

1. There exists a band of frequencies between $\sqrt{2C/M}$ and $\sqrt{2C/m}$ in which the wave-like solution does not exist. This is a characteristic feature of the elastic waves in the one-dimensional diatomic lattice. The existence of this band indicates that it is not possible to excite the vibrations of the frequencies lying within this band. Therefore, this band is known as the forbidden frequency gap. If we search for a solution in the forbidden frequency band with a real ω, then the wave vector $\vec{K}$ will be complex so that the wave is damped in space. The width of the forbidden frequency gap depends upon the masses m and

M of the two atoms. The first Brillouin zone extends from $k = -\pi/2a$ to $k = +\pi/2a$, so that the smallest possible wavelength is $4a$ corresponding to the value $k = +\pi/2a$ at the zone boundary.

2. If the two atoms are identical (i.e., $m = M$), then two branches meet at $k = \pm\pi/2a$as a consequence of the disappearance of the forbidden frequency gap. In such condition the frequency range for both mono-atomic and diatomic lattices are same and it is lies from 0 to $\sqrt{4C/m}$, i.e, $0 \le \omega^2 \le 4C/m$. For the mono-atomic lattice, the whole frequency range corresponds to the acoustical branch. On the other hand, it is divided into two parts for the diatomic lattice. The first part $0 \le \omega^2 \le 2C/m$ corresponds to the acoustical branch, while the second part $2C/m \le \omega^2 \le 4C/m$ corresponds to the optical branch.
3. When the mass of the heavier atom is increased, the value of $\sqrt{2C/M}$ decreases. Consequently, the acoustical branch yields in the downward direction and the optical branch tends to flatten. In the extreme situation (i.e, when $M \to \infty$), $(\omega_-)_{k=\pi/2a}$and $(\omega_+)_{k=0}$ are reduced to zero and $\sqrt{2C/m}$ respectively. Hence, the entire acoustical branch vanishes as shown in Figure 15.
4. The optical branch becomes flat completely and it is represented by a single frequency$\sqrt{2C/m}$. This physically means that all the atoms of lattice now become independent simple harmonic oscillators (i.e., they are absolutely free to oscillate independently with their natural frequency). In other words, the oscillations of each atom remain unaffected by the neighboring atoms. If we return from the extreme situation $M\to\infty$ to M, the separation between atoms ceases and again we observe the two different branches (i.e., the acoustical and the optical branches) as in the beginning. This case is equivalent to the case of the crystal formation. When the atoms come nearer to build a crystal then their energies levels merge with each other. Consequently, the broadening of all energies levels takes place in the form of an energy band. As soon as, the separation of atoms starts, the energy band is removed and each separated atom possesses its discrete energy levels.

5. When the mass of the lighter atom is decreased, the value of $\sqrt{2C/m}$ increases. Consequently, the optical branch shifts in the upward direction. In the extreme situation (i.e., when m → 0), the optical branch disappears completely. However, the acoustical branch remains unaffected. Hence, the diatomic lattice behaves like a mono-atomic lattice of lattice constant $2a$.

4.4. Modes of Vibrations in the Optical and the Acoustical Branches

From equations (38) and (39), we have:

$$\left|\frac{v_n}{u_n}\right| = \frac{v_0}{u_0} = \frac{2C-m\omega^2}{C(1+e^{-2ika})} \qquad (43i)$$

$$\left|\frac{v_n}{u_n}\right| = \frac{v_0}{u_0} = \frac{C(1+e^{2ika})}{2C-m\omega^2} \qquad (43ii)$$

For equation (43i) taking limits k → 0, we have

$$\frac{v_0}{u_0} = lim_{k\to 0}\frac{2C-m\omega^2}{C(1+e^{-2ika})} = \frac{2C-m(\omega^2)_{k\to 0}}{2C}$$

Using equation (42i) in the above equation for optical branch, we get

$$\left(\frac{v_0}{u_0}\right)_{\text{optical}} = \frac{2C-m(\omega_+^2)_{k\to 0}}{2C} = \frac{2C-m.2C\left(\frac{1}{m}+\frac{1}{M}\right)}{2C} = \frac{2C\left\{1-1-\frac{m}{M}\right\}}{2C} = -\frac{m}{M}$$

Thus, if we consider the optical modes of vibrations, then we find that the two atoms of mass m and M oscillate in the opposite directions [see Figure 16]. For the optical modes of vibrations, the amplitudes of two atoms are inversely proportional to their masses.

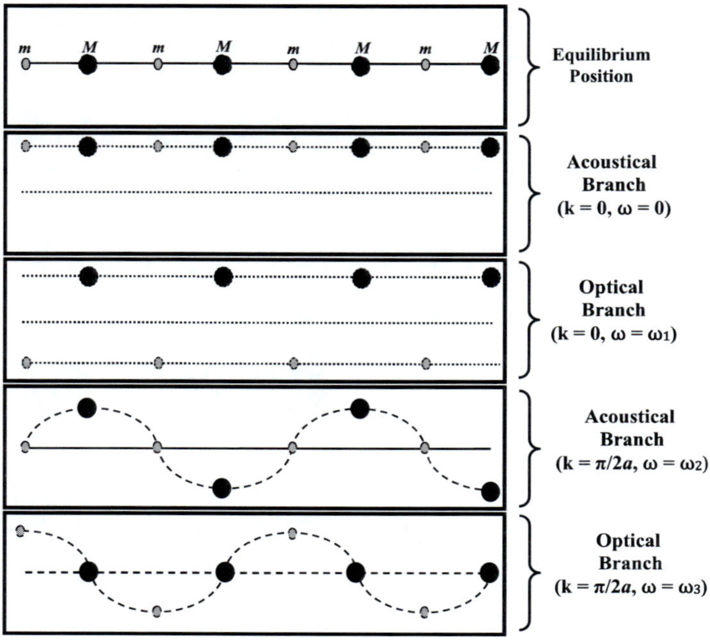

Figure 16. Vibrations of one-dimensional diatomic chain in optical and acoustical branches under different conditions.

Using equation (42*i*) in above equation for acoustical branch, we get

$$\left(\frac{v_0}{u_0}\right)_{\text{acoustical}} = \frac{2C - m.(\omega_-^2)_{k\to 0}}{2C} = \frac{2C - m.0}{2C} = \frac{2C}{2C} = 1$$

Thus, if we consider the acoustical modes of vibrations, then we find that the two atoms of mass *m* and *M* oscillate in the same directions [see Figure 16]. For the acoustical modes of vibrations, the amplitudes of two atoms are same.

$$\therefore \left(\frac{v_0}{u_0}\right)_{k=0} = \begin{Bmatrix} -\frac{m}{M}\text{for optical branch} \\ 1 \text{ for acoustical branch} \end{Bmatrix} \quad (44i)$$

For equation (43*i*) and (43*ii*) taking limits k → π/2*a* and keeping in mind equation (42*ii*), we have

$$\left(\frac{v_0}{u_0}\right)_{\text{optical}} = lim_{k\to\pi/2a}\frac{2C-m(\omega_+^2)_{k\to\pi/2a}}{C(1+e^{-2ika})} = \frac{2C-m\frac{2C}{m}}{C(1+e^{-2ika})} = 0$$

$$\left(\frac{v_0}{u_0}\right)_{\text{acoustical}} = lim_{k\to\pi/2a}\frac{C(1+e^{2ika})}{2C-M(\omega_-^2)_{k\to\pi/2a}} = \frac{C(1+e^{2ika})}{2C-M\frac{2C}{M}} = \infty$$

$$\therefore \left(\frac{v_0}{u_0}\right)_{k=\pi/2a} = \begin{Bmatrix} 0 \text{ for optical branch} \\ \infty \text{ for acoustical branch} \end{Bmatrix} \qquad (44ii)$$

From Figure 16, it is clear that the heavier and the lighter atoms of masses M and m respectively are initially at rest in their equilibrium position. They oscillate with equal amplitudes and zero phase difference in the limiting case k → 0 for the acoustical branch. It means that both kinds of atoms are displaced by the same amount in the same direction. Thus, the wavelength is infinity and frequency is zero for this case. Consequently, the entire lattice is displaced.

On the other hand, the heavier and the lighter atoms oscillate with equal amplitudes and 180° phase difference in the limiting case k → 0 for the optical branch. It means that both kinds of atoms are again displaced in opposite directions. The center of mass of two neighboring atoms (one is heavy atom and other is lighter) remains stationary [since $u_0.m + v_0.M = 0$]. The wavelength of the wave is infinity as a signature of the same displacement of all the atoms of the same kind at all times. This condition is equivalent to the internal motion of the atoms in the unit cell. An ionic chain (i.e., a linear arrangement of oppositely charged ions placed alternately) is a convenient example of such a motion where the oppositely charged ions oscillate out of phase under the influence of an electromagnetic wave.

From the dispersion curve shown in Figure 14, it is clear that the acoustical branch has minimum frequency ($\omega_{min} = 0$) at $k = 0$, while the optical branch has maximum frequency($\omega_{max} = \sqrt{2C(M+m)/mM}$)= ω_1 (say) at $k = 0$. Further, the acoustical and the optical branches terminate at frequencies $\sqrt{2C/M} = \omega_2$ (say) and $\sqrt{2C/m} = \omega_3$(say) respectively at the boundary of the first Brillouin zone (i.e., k = $\pi/2a$). Thus, the waves corresponding to the frequencies $\omega_2<\omega<\omega_3$ and $\omega>\omega_1$ do not propagate through the crystal lattice. Since $\omega_2<\omega_3$, therefore both branches do not cross each other and the chain acts like a band-pass filter.

The plot of amplitude ratio (v_0/u_0) versus k is shown in Figure 17.

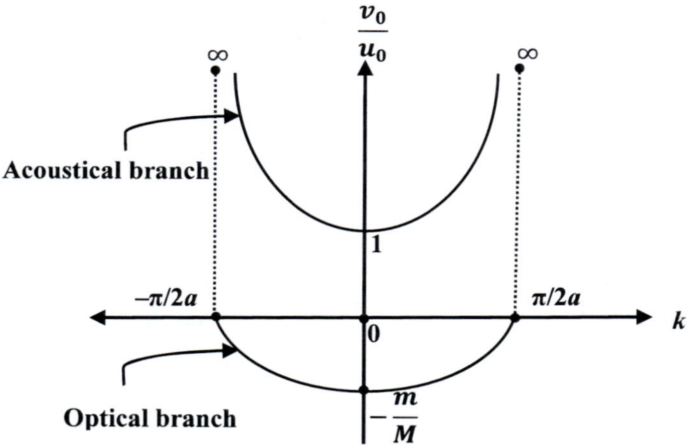

Figure 17. Plot showing the variation of the amplitude ratio (v_0/u_0) as a function of k for optical and acoustical branches.

From Figure 17 we see that the ratio v_0/u_0 of the amplitudes tends to infinity for the acoustics branch at the boundary (i.e., $k = \pm\pi/2a$) of the first Brillouin zone, while this ratio reduces to zero for the optical branch. Corresponding to the acoustical branch, the atoms oscillate in the same phase with the same amplitude at $k = 0$. As k increases from 0 to π/2a, the wavelength is decreased. Consequently, the amplitude of the lighter atoms also decreases and it is reduced to zero at Brillouin zone boundary and hence only the heavier atoms continue the oscillations. This means that only one kind of atoms (i.e., either heavier atoms or the lighter atoms) oscillates at the Brillouin zone boundary.

5. Lattice Vibrations of a One-Dimensional Diatomic Chain Having Identical Atoms of Same Masses Attached with Springs of Different Force Constants

Let us consider the vibrations of a one-dimensional diatomic lattice of lattice constant $2a$. Suppose that this one-dimensional diatomic chain has two identical atoms of mass m per primitive cell. The atoms in the array are linked alternately with the springs of two different strengths such that their force constants between the nearest-neighbor atoms are alternately $C_1$and C_2 ($C_1>C_2$) as shown in Figure 18. Each atom is separated by its nearest neighbors by a distance a.If we apply the harmonic approximation and consider the nearest neighbor interaction only as we have discussed in section 4.1, then potential energy of the whole system may be expressed as:

$$V = V_0 + \frac{1}{2}\sum_n\{C_2(u_{n+1} - v_n)^2 + C_1(v_n - u_n)^2 + C_2(u_n - v_{n-1})^2\} \tag{45}$$

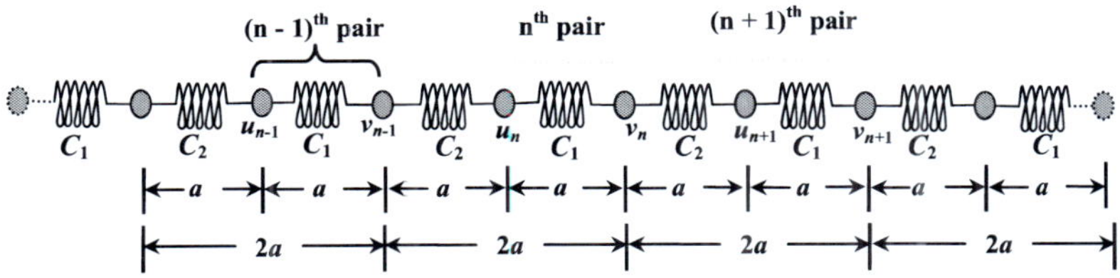

Figure 18. Lattice vibrations of a one-dimensional diatomic chain for identical atoms of mass m connected with the springs of alternate force constants $C_1$and C_2.

Here u_n and v_n are the instantaneous displacements of atoms in nth pair. The equations of motions of atoms nth pair can be, therefore, expressed as:

$$m\frac{\partial^2 u_n}{\partial t^2} = -\frac{\partial}{\partial u_n}\left[V_0 + \frac{1}{2}\sum_n\{C_2(u_{n+1} - v_n)^2 + C_1(v_n - u_n)^2 + C_2(u_n - v_{n-1})^2\}\right]$$

$$m\frac{\partial^2 u_n}{\partial t^2} = -\frac{\partial}{\partial v_n}\Big[V_0 + \tfrac{1}{2}\Sigma_n\{C_2(u_{n+1} - v_n)^2 + C_1(v_n - u_n)^2 +$$
$$C_2(u_n - v_{n-1})^2\}\Big]$$

Solving these equations, we obtain:

$$m\frac{\partial^2 u_n}{\partial t^2} = C_2 v_{n-1} + C_1 v_n - (C_1 + C_2)u_n \qquad (46i)$$

$$m\frac{\partial^2 v_n}{\partial t^2} = C_2 u_{n+1} + C_1 u - (C_1 + C_2)v_n \qquad (46ii)$$

Consider the general solution (38*i*) for (46*i*), we have

$$m(-\omega^2 u_n) = (C_1 + C_2 e^{-2ika})v_n - (C_1 + C_2)u_n$$

or

$$\{(C_1 + C_2) - m\omega^2\}u_n - \left(C_1 + C_2 e^{-2ika}\right)v_n = 0 \qquad (47i)$$

Consider the general solution (38*ii*) for (46*ii*), we have:

$$m(-\omega^2 v_n) = \left(C_1 + C_2 e^{2ika}\right)u_n - (C_1 + C_2)v_n$$

or

$$\left(C_1 + C_2 e^{2ika}\right)u_n - \{(C_1 + C_2) - m\omega^2\}v_n = 0 \qquad (47ii)$$

Solving equations (47*i*) in (47*ii*) using determinant method, we have:

$$\begin{vmatrix} \{(C_1 + C_2) - m\omega^2\} & -(C_1 + C_2 e^{-2ika}) \\ \left(C_1 + C_2 e^{2ika}\right) & -\{(C_1 + C_2) - m\omega^2\} \end{vmatrix} = 0$$

i.e.,

$$-\{(C_1 + C_2) - m\omega^2\}^2 + \left(C_1 + C_2 e^{2ika}\right).\left(C_1 + C_2 e^{-2ika}\right) = 0$$

i.e.,

$$-(C_1+C_2)^2 - m^2\omega^4 + 2m\omega^2(C_1+C_2) + C_1^2 + C_2^2 + C_1C_2\left(e^{2ika} + e^{-2ika}\right) = 0$$

i.e.,

$$-2C_1C_2 - m^2\omega^4 + 2m\omega^2(C_1+C_2) + C_1C_2\,.\,2\cos(2ka) = 0$$

i.e.,

$$m^2\omega^4 - 2m\omega^2(C_1+C_2) + 2C_1C_2.\{1-\cos(2ka)\} = 0$$

i.e.,

$$m^2\omega^4 - 2m\omega^2(C_1+C_2) + 2C_1C_2.\{2\,sin^2ka\} = 0$$

i.e.,

$$m^2\omega^4 - 2m\omega^2(C_1+C_2) + 4C_1C_2.sin^2ka = 0$$

$$\therefore \omega^4 - \frac{2(C_1+C_2)}{m}.\omega^2 + \frac{4C_1C_2}{m^2}.sin^2ka = 0 \qquad (48)$$

The equation (48) represents the dispersion relation correlating ω with k for the lattice vibrations of one-dimensional diatomic chain. As this relation is a quadratic equation of ω^2. Therefore the roots of ω^2 can be expressed as:

$$\omega^2 = \frac{\frac{2(C_1+C_2)}{m} \pm \left[\frac{4(C_1+C_2)^2}{m^2} - \frac{16C_1C_2}{m^2}sin^2ka\right]^{1/2}}{2}$$

or

$$\omega^2 = \frac{(C_1+C_2)}{m} \pm \left[\frac{(C_1+C_2)^2}{m^2} - \frac{4C_1C_2}{m^2}sin^2ka\right]^{1/2} \qquad (49)$$

Now following two cases arise:

Case (i) When $k = 0$, i.e., sin $ka = 0$, then equation (49) reduces to:

$$\omega^2 = \frac{(C_1+C_2)}{m} \pm \frac{(C_1+C_2)}{m}$$

i.e.,

$$\omega^2 = \begin{cases} \frac{(C_1+C_2)}{m} + \frac{(C_1+C_2)}{m}; \text{ for positive sign} \\ \frac{(C_1+C_2)}{m} - \frac{(C_1+C_2)}{m}; \text{ for negative sign} \end{cases}$$

i.e.,

$$\omega = \begin{Bmatrix} \sqrt{\frac{2(C_1+C_2)}{m}} = (\omega_+)_{k=0}; \text{for positive sign} \\ 0 = (\omega_-)_{k=0}; \text{ for negative sign} \end{Bmatrix} \tag{50i}$$

Case (ii) When $k = \pi/2a$, i.e., $\sin ka = 1$, then equation (49) reduces to:

$$\omega^2 = \frac{(C_1+C_2)}{m} \pm \left[\frac{(C_1+C_2)^2}{m^2} - \frac{4C_1C_2}{m^2}\right]^{1/2}$$

i.e.,

$$\omega^2 = \frac{(C_1+C_2)}{m} \pm \frac{(C_1-C_2)}{m}$$

i.e.,

$$\omega^2 = \begin{Bmatrix} \frac{(C_1+C_2)}{m} + \frac{(C_1-C_2)}{m}; \text{for positive sign} \\ \frac{(C_1+C_2)}{m} - \frac{(C_1-C_2)}{m}; \text{ for negative sign} \end{Bmatrix}$$

i.e.,

$$\omega = \begin{Bmatrix} \sqrt{\frac{2C_1}{m}} = (\omega_+)_{k=\pi/2a}; \text{for positive sign} \\ \sqrt{\frac{2C_2}{m}} = (\omega_-)_{k=\pi/2a}; \text{ for negative sign} \end{Bmatrix} \tag{50ii}$$

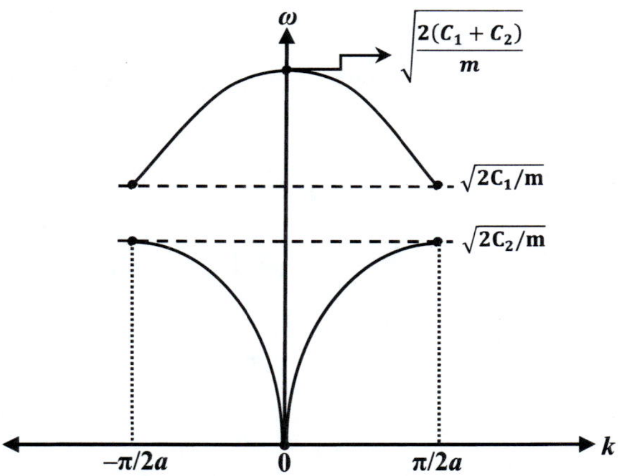

Figure 19. Dispersion curve of a one-dimensional diatomic chain for identical atoms of mass m connected with the springs of alternate force constants C_1 and C_2.

Thus, the corresponding dispersion curve (i.e., plot of ω against k) consists of two branches as shown in Figure 19.

Now, the forbidden frequency gap lies in the frequency range from $\sqrt{2C_2/\mathrm{m}}$ to $\sqrt{2C_1/\mathrm{m}}$ and its width depends upon the mass m of the atoms and theforce constants C_1 and C_2 of the two kinds of alternate springs. If the two springs are identical (i.e., $C_1 = C_2$), then two branches will meet at $k = \pm\pi/2a$ and the forbidden frequency gap will be disappeared.

6. Lattice Vibrations of Two-Dimensional Atomic Chain

Let us consider the case of a two-dimensional square lattice having lattice constant a to understand the lattice vibrations of two-dimensional atomic chain. Suppose that this square lattice has identical atoms each of mass m. For the convenience of mathematical calculations, let us assume the nearest neighbor interaction of each atom only with its neighboring atoms located at distance a i.e., we are not considering the neighboring atoms located at diagonal positions.

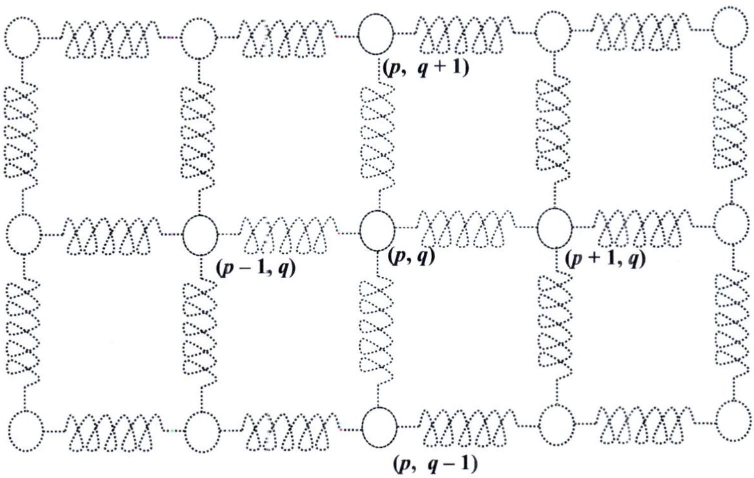

Figure 20. Lattice vibrations of a two-dimensional atomic chain.

Thus, if we consider an atom located at a position marked by (p, q) as shown in the Figure 20, then the positions of its nearest neighboring atoms will be ($p \pm 1$, q) and (p, $q \pm 1$). Keeping in mind the harmonic approximation the potential energy of the whole two-dimensional square lattice can be expressed as:

$$V = V_0 + \frac{C}{2}\sum_{p,q}\left\{\left(u_{p+1,q} - u_{p,q}\right)^2 + \left(u_{p,q} - u_{p-1,q}\right)^2 + \left(u_{p,q+1} - u_{p,q}\right)^2 + \left(u_{p,q} - u_{p,q-1}\right)^2\right\}$$

If $u_{p,q}$ is the instantaneous displacement of the atom located at the position (p, q) during harmonic motion, then its equation of motion is:

$$m\frac{\partial^2 u_{p,q}}{\partial t^2} = -\frac{\partial V}{\partial u_{p,q}}$$

Substituting the value of V in the equation of motion of the atom (p, q), we have:

$$m\frac{\partial^2 u_{p,q}}{\partial t^2} = -\frac{C}{2}\frac{\partial}{\partial u_{p,q}}\left\{\left(u_{p+1,q} - u_{p,q}\right)^2 + \left(u_{p,q} - u_{p-1,q}\right)^2 + \left(u_{p,q+1} - u_{p,q}\right)^2 + \left(u_{p,q} - u_{p,q-1}\right)^2\right\}$$

$$m\frac{\partial^2 u_{p,q}}{\partial t^2} = -\frac{C}{2}2.\left\{\left(u_{p,q} - u_{p-1,q}\right) - \left(u_{p+1,q} - u_{p,q}\right) + \left(u_{p,q} - u_{p,q-1}\right) - \left(u_{p,q+1} - u_{p,q}\right)\right\}$$

$$m\frac{\partial^2 u_{p,q}}{\partial t^2} = C.\left\{u_{p+1,q} + u_{p-1,q} + u_{p,q+1} + u_{p,q-1} - 4u_{p,q}\right\} \quad (51)$$

Suppose that the general solution of the equation (51) is:

$$u_{p,q} = u_0.\exp\{-i(\omega t - k_x.pa - k_y.qa)\} = u_0 e^{-i\omega t} e^{ik_x.pa} e^{ik_y.qa} \quad (52)$$

i.e., $u_{p\pm1,q} = e^{\pm ik_x a} u_{p,q}$ and $u_{p,q\pm1} = e^{\pm ik_y a} u_{p,q}$

Using the solution (52) in the equation (51), we have:

$$-m\omega^2 u_{p,q} = C\left[\left(e^{ik_x a} + e^{-ik_x a}\right) + \left(e^{ik_y a} + e^{-ik_y a}\right) - 4\right].u_{p,q}$$

$$\text{or} -m\omega^2 = C\left[2\cos k_x a + 2\cos k_y a - 4\right]$$

$$\therefore \omega^2 = \frac{2C}{m}\left[2 - \cos k_x a - \cos k_y a\right] \qquad (53)$$

The equation (53) is the dispersion relation for the lattice vibrations of two-dimensional square lattice.

Using cosine half angle formula, we get another form of the dispersion relation as:

$$\omega^2 = \frac{4C}{m}\left[sin^2\frac{k_x a}{2} + sin^2\frac{k_y a}{2}\right] \qquad (54)$$

The corresponding first Brillouin zone is defined as follows:

$$-\frac{\pi}{a} \le k_x \le \frac{\pi}{a};\ -\frac{\pi}{a} \le k_y \le \frac{\pi}{a}$$

7. LATTICE VIBRATIONS OF A THREE-DIMENSIONAL ATOMIC CHAIN

Now we think about a simple cubic crystal to understand the lattice vibrations of a three-dimensional atomic chain. Thus, we can consider the case of a three-dimensional cubic lattice having lattice constant *a*.

Suppose that this cubic lattice has identical atoms each of mass *m*. For the simple mathematical analysis, we consider only the nearest neighbor interaction of each atom with its neighboring atoms located at distance *a* i.e., we are not considering the neighboring atoms located at diagonal positions.

Thus, if we consider an atom located at a position marked by (p, q, r) as shown in Figure 21, then the positions of its nearest neighboring atoms will be $(p \pm 1, q, r)$, $(p, q \pm 1, r)$ and $(p, q, r \pm 1)$.

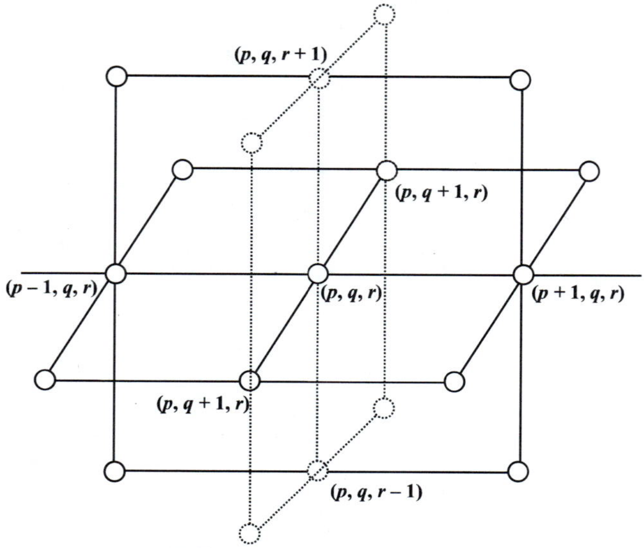

Figure 21. Lattice vibrations of a three-dimensional atomic chain.

Keeping in mind the harmonic approximation the potential energy of the whole three-dimensional cubic lattice can be expressed as:

$$V = V_0 + \frac{C}{2}\sum_{p,q,r}\Big\{\left(u_{p+1,q,r} - u_{p,q,r}\right)^2 + \left(u_{p,q,r} - u_{p-1,q,r}\right)^2 + \left(u_{p,q+1,r} - u_{p,q,r}\right)^2 + \left(u_{p,q,r} - u_{p,q-1,r}\right)^2 + \left(u_{p,q,r+1} - u_{p,q,r}\right)^2 + \left(u_{p,q,r} - u_{p,q,r-1}\right)^2\Big\}$$

If $u_{p,q,r}$ is the instantaneous displacement of the atom located at the position (p, q, r) during harmonic motion, then its equation of motion is:

$$m\frac{\partial^2 u_{p,q,r}}{\partial t^2} = -\frac{\partial V}{\partial u_{p,q,r}}$$

Substituting the value of V in the equation of motion of the atom (p, q, r), we have:

$$m\frac{\partial^2 u_{p,q,r}}{\partial t^2} = -\frac{C}{2}\frac{\partial}{\partial u_{p,q,r}}\left\{\left(u_{p+1,q,r} - u_{p,q,r}\right)^2 + \left(u_{p,q,r} - u_{p-1,q,r}\right)^2 + \left(u_{p,q+1,r} - u_{p,q,r}\right)^2 + \left(u_{p,q,r} - u_{p,q-1,r}\right)^2 + \left(u_{p,q,r+1} - u_{p,q,r}\right)^2 + \left(u_{p,q,r} - u_{p,q,r-1}\right)^2\right\}$$

$$m\frac{\partial^2 u_{p,q,r}}{\partial t^2} = -\frac{C}{2}2.\left\{\left(u_{p,q,r} - u_{p-1,q,r}\right) - \left(u_{p+1,q,r} - u_{p,q,r}\right) + \left(u_{p,q,r} - u_{p,q-1,r}\right) - \left(u_{p,q+1,r} - u_{p,q,r}\right) + \left(u_{p,q,r} - u_{p,q,r-1}\right) - \left(u_{p,q,r+1} - u_{p,q,r}\right)\right\}$$

$$m\frac{\partial^2 u_{p,q,r}}{\partial t^2} = C.\left\{u_{p+1,q,r} + u_{p-1,q,r} + u_{p,q+1,r} + u_{p,q-1,r} + u_{p,q,r+1} + u_{p,q,r-1} - 6u_{p,q,r}\right\} \tag{55}$$

Suppose that the general solution of the equation (55) is:

$$u_{p,q,r} = u_0.\exp\left\{-i\left(\omega t - k_x.pa - k_y.qa - k_z.ra\right)\right\} = u_0 e^{-i\omega t} e^{ik_x.pa} e^{ik_y.qa} e^{ik_z.ra} \tag{56}$$

In light of equation (56), we can write:

$$u_{p\pm1,q,r} = e^{\pm ik_x a} u_{p,q,r}$$

$$u_{p,q\pm1,r} = e^{\pm ik_y a} u_{p,q,r}$$

$$u_{p,q,r\pm1} = e^{\pm ik_z a} u_{p,q,r}$$

Using the solution (56) in the equation (55), we have:

$$-m\omega^2 u_{p,q,r} = C\left[\left(e^{ik_x a} + e^{-ik_x a}\right) + \left(e^{ik_y a} + e^{-ik_y a}\right) + \left(e^{ik_z a} + e^{-ik_z a}\right) - 6\right].u_{p,q}$$

or

$$-m\omega^2 = C[2\cos k_x a + 2\cos k_y a + 2\cos k_z a - 6]$$
$$\therefore \omega^2 = \frac{2C}{m}[3 - \cos k_x a - \cos k_y a - \cos k_z a] \quad (57)$$

The equation (57) is the dispersion relation for the lattice vibrations of three-dimensional cubic crystal lattice.

Using cosine half angle formula, we get another form of the dispersion relation as:

$$\omega = \omega_0 \left[sin^2 \frac{k_x a}{2} + sin^2 \frac{k_y a}{2} + sin^2 \frac{k_z a}{2}\right]^{1/2} \quad (58)$$

Here $\omega_0 = \sqrt{\frac{4C}{m}}$ is constant for the cubic crystal under consideration.

8. CONCEPT OF PHONONS

The dispersion relations discussed in the above sections for the lattice vibrations are based on the theories of the simple harmonic oscillator, ball-spring model and the Hooke's law. These theories follow the classical approach and each theory is based on classical mechanics.

In 1907, Einstein proposed a theory of heat capacity by using the Planck's quantum hypothesis. According to Einstein's theory, all atoms of a solid vibrate about their rest positions such that we can treat the solid as an assembly of simple harmonic oscillators. In the quantum mechanics, the energy levels of a simple harmonic oscillator are given by the expression:

$$E_n = \left(n + \frac{1}{2}\right)\hbar\omega \quad (59)$$

Here n = 0, 1, 2, 3, ...

The corresponding energy levels can be obtained by putting the values of *n* as shown in the Figure 22.

In the same manner, the energy levels of a mono-atomic chain having N unit cells and one atom per unit cell can be expressed as:

$$E_n = \left(n + \frac{1}{2}\right) \hbar\omega(k) \tag{60}$$

Here $k = 0, \frac{2\pi}{Na}, 2\frac{2\pi}{Na}, 3\frac{2\pi}{Na}, 4\frac{2\pi}{Na}, \ldots\ldots\ldots\ldots\ldots\ldots$

The corresponding energy levels are shown in Figure 22.

Thus, each lattice vibration mode is quantized in the framework of quantum mechanics and it can be treated as characteristics feature of the quantum harmonic oscillator having frequency and wave vector $\vec{k}$. This quantized lattice vibration mode signifies the concept of phonon. Phonons are not real particles like atoms or electrons. The number of phonons is related to vibration energy. The number would increase when the temperature increases. A phonon is the quantum of sound similar to the photon that is the quantum of electromagnetic radiation.

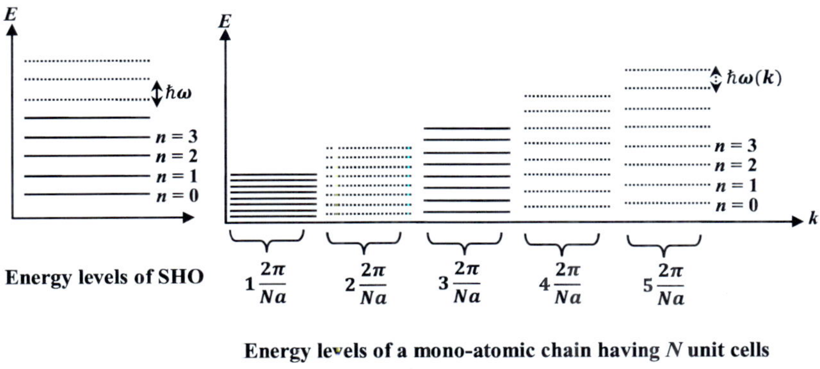

Figure 22. Analogy between the energy levels of a SHO and a mono-atomic chain.

Thus, we choose the concept of the simple harmonic oscillator as a fascinating model due to the organization of its energy levels. Therefore, this model can be used to explain the lattice dynamics of the crystals and its quantum-mechanical explanation in terms of the phonons.

Acknowledgments

I express distinctive gratitude to my wife Dr. Shipra Saraswat for her constant inspiration, blessings and faith in me. I also thank my adorable son Master Swastik for his sacrifices during the course of this work. I am deeply thankful to my research scholar Mr. Shiv Kumar Pal, Physics Department, B.H.U., Varanasi, India who participated thoroughly in the process of the proofreading of this chapter. Last but not least, I would like to thank my parents and almighty for their endless blessings and kindness.

REFERENCES

[1] Kosevich, A. M. (2005). *The Crystal Lattice: Phonons, Solitons, Dislocations, Superlattices.* Wiley-Vch Verlag GmbH & Co. KGaA.

[2] Kittel, C. (2004). *Introduction to Solid State Physics*, Chapter 4, John Wiley & Sons, Inc.

[3] Patterson, J. and Bailey, B. (2007). *Solid-State Physics: Introduction to the Theory,* Springer-Verlag Berlin Heidelberg.

[4] Ashcroft, N. W. and Mermin, N. D. (1976). *Solid State Physics,* Holt, Rinehart and Winston Publisher.

ABOUT THE AUTHOR

Dr. Neeraj Mehta is the associate professor in the Department of Physics, Banaras Hindu University, Varanasi, India. He obtained his MSc degree in Physics with the third position in Merit from the C.S.J.M. University, Kanpur,

India and completed his PhD from the Harcourt Butler Technological Institute, Kanpur, India under the supervision of Prof. Ashok Kumar.

While pursuing his PhD, he earned the prestigious Junior Research Fellowship (JRF) and Senior Research Fellowship (SRF) by qualifying National Eligibility Test (NET). He has published more than 200 research papers in different reputed national and international journals of Elsevier, Springer, Institute of Physics (IOP), American Chemical Society (ACS), Royal Society of Chemistry (RSC), Taylor & Francis, American Scientific Publishers (ASP) and attended various national and international conferences. During his 15 years of teaching experience, Dr. Mehta has authored several books for the students of undergraduate courses. He is also a referee for various journals of international repute. He is a life member of the Indian Association of Physics Teachers (IAPT), Indian Physics Association (IPA) and International Forum of Chalcogeniders (IFC). He has produced four PhDs and completed four major projects. He is the member of the editorial boards of Heliyon (An Elsevier journal) and Materials Focus (A reputed journal of American Scientific Publishers).

About the Editor

Yilun Shang, PhD
Associate Professor
Northumbria University, Newcastle upon Tyne, UK
Email: shylmath@hotmail.com

Yilun Shang is an Associate Professor with Northumbria University. He received PhD degrees in mathematics from Shanghai Jiao Tong University in 2010. His research interests include complex networks, complex systems, applied probability, combinatorics, algorithms and computation. He is the recipient of 2016 Dimitrie Pompeiu Prize and is on the Editorial Boards of several international journals including *Scientific Reports (Nature)*, *European Journal of Pure and Applied Mathematics*, *IEEE Access*, *KSII Transactions on Internet and Information Systems*.

INDEX

E

F

G

H

I

L

M

N

O